NYC

Department of Education

BUREAU OF NONPUBLIC SCHOOL REIMBURSABLE SERVICES

Title I
Elementary School Activities

with **THE GEOMETER'S**
SKETCHPAD®
VERSION 5

Key Curriculum Press
INNOVATORS IN MATHEMATICS EDUCATION

Writers:	Dan Bennett, Christopher Casey, Greg Clarke, Larry Copes, Deidre Grevious, Lynn Hughes, Rhea Irvine, Ross Isenegger, Nick Jackiw, Tobias Jaw, Amy Lamb, Paul Kunkel, Ann Lawrence, Andres Marti, Daniel Scher, Nathalie Sinclair, Scott Steketee, Kelly Stewart, Kevin Thompson
Reviewers:	Karen Anders, Susan Beal, Janet Beissinger, Dudley Brooks, Greg Clarke, Gord Cooke, Larry Copes, Judy Dussiaume, Paul Gautreau, Shawn Godin, Paul Goldenberg, Lynn Hughes, Scott Immel, Ross Isenegger, Sarah Kasten, Cathy Kelso, Amy Lamb, Dan Lufkin, Aaron Madrigal, Linda Modica, Margo Nanny, Henry Picciotto, Nicolle Rosenblatt, Joan Scher, Dick Stanley, Tom Steinke, Glenda Stewart, John Threlkeld, Philip Wagreich, Ken Waller, Bill Zahner, Danny Zhu
Field Testers:	Laura Adler, Joëlle Auberson, Vera Balarin, Kim Beames, Judy Bieze, Caron Cesa, Heather Darby, Robin Glass, Cherish Hansen, Layne Hudes, Lynn Hughes, Scott Immel, Susan Friedman, Ann Lawrence, Vanessa Mamikunian, Michelle Mancini, Michelle Moore, Margo Nanny, Kelly O'Keefe, Tina Pierorazio, Leslie Profeta, Dechelle Rasheed, Cheryl Schafer, Joan Scher, Kimberly Scheier, JoAnne Searle, Char Soucy, Jessie Starr, Ruth Steinberg, Nancy Stevenson, Terry Suetterlein, Mona Sussman, William Vaughn, Ethan Weker, Angie Whaley
Module Editors:	Elizabeth DeCarli, Karen Greenhaus
Activity Editors:	Rhea Irvine, Daniel Scher, Josephine Noah, Scott Steketee, Cindy Clements, Joan Lewis, Silvia Llamas-Flores, Kendra Lockman, Lenore Parens, Glenda Stewart, Kelly Stewart
Production Editors:	Angela Chen, Andrew Jones, Christine Osborne
Other Contributers:	Judy Anderson, Elizabeth Ball, Tamar Chestnut, Brady Golden, Ashley Kuhre, Nina Mamikunian, Marilyn Perry, Emily Reed, Ann Rothenbuhler, Juliana Tringali, Jeff Williams
Copyeditor:	Jill Pellarin
Printer:	Lightning Source, Inc.
Executive Editor:	Josephine Noah
Publisher:	Steven Rasmussen

Key Curriculum Press
1150 65th Street
Emeryville, CA 94608
510-595-7000
editorial@keypress.com
www.keypress.com

ISBN: 978-1-60440-244-5
10 9 8 7 6 5 4 3 2 1   15 14 13 12 11

Contents

Chapter 5: Grade 5 Activities

Sketchpad Resources

Sketchpad Learning Center

The Learning Center provides a variety of resources to help you learn how to use Sketchpad, including overview and classroom videos, tutorials, Sketchpad Tips, sample activities, and links to online resources. You can access the Learning Center through Sketchpad's start-up screen or through the Help menu.

The Learning Center has three main sections:

Welcome Videos

These videos introduce Sketchpad from the point of view of students and teachers, and give an overview of the big ideas and new features of Sketchpad 5.

Using Sketchpad

This section includes 12 self-guided tutorials with embedded videos, 70 Sketchpad Tips, and links to local and online resources.

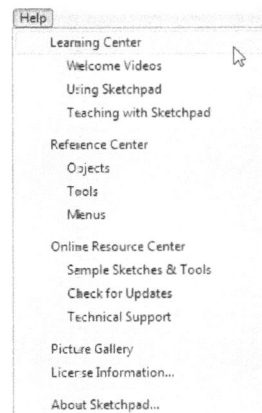

Using Sketchpad

Getting Started Tutorials
Learn how to use Sketchpad while exploring central mathematical ideas. Follow step-by-step instructions supported by online videos.

Sketchpad Tips
Explore dozens of tips—provided as both comic strips and short online videos—that give you a quick overview of Sketchpad's tools and menu commands.

Reference Center
Browse, search by topic, or use the index to find the information you need in this comprehensive reference manual. You can also access the Reference Center directly from Sketchpad's Help menu.

Online Resource Center
Consult an extensive collection of free resources for Sketchpad users, including sample sketches and tools, an advanced sketch gallery, and information about software updates. You can also access the Online Resource Center directly from Sketchpad's Help menu.

Teaching with Sketchpad

This section includes videos and articles describing how teachers make effective use of Sketchpad and how it affects their students' attitudes and mathematical understanding. There are over 40 sample activities, each with an overview, teaching notes, student worksheet, and sketches, that you can use with students to support your subject area, level, and curriculum.

Other Sketchpad Resources

Sketchpad contains resources for beginning and advanced users.

- **Reference Center:** This digital resource, which is accessed through the Help menu, is the complete reference manual for Sketchpad, with detailed information on every object, tool, and menu command. The Reference Center includes a number of How-To sections, an index, and full-text search capability.

- **Online Resource Center:** The Geometer's Sketchpad Resource Center (www.dynamicgeometry.com) contains many sample sketches and advanced toolkits, links to other Sketchpad sites, technical information (including updates and frequently asked questions), and detailed documentation for JavaSketchpad, which allows you to embed dynamic constructions in a web page.

- **Sketch Exchange:** The Sketchpad Sketch Exchange™ (sketchexchange.keypress.com) is a community site where teachers share sketches and other resources with Sketchpad users. Browse by keyword or topic for sketches that interest you, or ask questions and share ideas in the forum.

- **Sample Sketches & Tools:** You can access many sketches, including some with custom tools, through Sketchpad's Help menu. You can use some sample sketches as demonstrations, others to get tips and information about particular constructions, and others to access custom tools that you can use to perform special constructions. These sketches are also available under General Resources at the Sketchpad Resource Center (www.dynamicgeometry.com).

- **Online Courses:** Key Curriculum Press offers moderated online courses that last six weeks, allowing you to immerse yourself in learning how to use Sketchpad in your teaching. For more information, see Sketchpad's Learning Center, or go to www.keypress.com/onlinecourses.

- **Other Professional Development:** Key Curriculum Press offers free webinars on a regular basis. You can also arrange for one-day or three-day face-to-face workshops for your district or school. For more information, go to www.keypress.com/pd.

Addressing Grade-Level Learning Objectives

The table below shows how the activities in this collection align to the Title I Mathematics Learning Objectives for Grades 1–5.

The following Learning Objectives are not listed in the table, because virtually every activity addresses them:

- Use tools, such as manipulatives or sketches, to model/explain problems or justify solutions (Grades 1 and 2)

- Appropriately use, discuss, and explain mathematical terms and ideas in verbal or written form (Grades 1 and 2)

- Express the solution of a problem clearly and logically by using the appropriate mathematical notation or terminology; support solutions with evidence in verbal or written form (Grades 3–5)

The Activity Notes for each activity guide you in using dynamic models, manipulatives, and constructions to model and explore mathematical concepts with your students, in guiding class discussions, and in helping students explain and justify their solutions using appropriate mathematical terminology.

Activity Title	Learning Objective
Grade 1	
Skip-Counting: Patterns on the Hundreds Chart	Count, read, and write whole numbers to 100
	Skip count by 2s, 5s, and 10s to 100
Place-Value Counter: Ones, Tens, Hundreds, Thousands	Explore place value by bundling objects into groups of tens and ones (24 = 2 tens + 4 ones or 24 ones)
Comparing Heights: Logic and Deduction	Represent and compare data (largest, smallest, most often, least often) using pictures, picture graphs, bar graphs and tally charts
Grade 2	
Circles and Squares: Representing an Unknown	Construct and solve simple arithmetic open sentences involving an unknown value in any position ($\square + 17 = 39$; $39 - \square = 17$)
Bugs in Groups: Dividing into Groups of Equal Size	Use repeated addition, arrays, and counting by multiples to develop readiness for multiplication
	Use repeated subtraction, equal sharing, and forming equal groups with remainders to develop readiness for division
How Much Is Half: Size of the Unit Whole	Recognize, name, order, and compare unit fractions (1/4, 1/2, 3/4)
Dynamic Triangles: Attributes of Triangles Dynamic Rectangles: Attributes of Rectangles Square or Not: Properties of Squares Circles All Around: Parts of a Circle	Name, identify, describe, draw, and compare two-dimensional shapes (circle, triangle, square, etc.) and three-dimensional figures (sphere, pyramid, cube, etc.)

Activity Title	Learning Objective
Grade 3	
Place-Value Counter: Get to the Target	Identify the place value for each digit in whole numbers to 10,000; use this understanding to round whole numbers to the nearest 10, 100, and 1,000
Factor Puzzles: Number Sense and Logical Reasoning Jump Along: Multiplication on the Number Line Jump Along: Factor Families on the Number Line	Know multiplication and related division facts through 10×10 Use the inverse relationship of multiplication and division to compute and check results ($6 \times 5 = 30$; $30 \div 5 = 6$)
Comparing Fractions: Number Sense and Benchmarks Dividing and Subdividing: Fractions on the Number Line Jump Along: Equivalent Fractions on the Number Line	Compare and order unit fractions (1/2, 1/3, 1/4, 1/5, 1/6, and 1/10) and find their approximate locations on a number line Explore equivalent fractions using manipulatives (4/6 = 2/3)
Missing Pieces: Polygons that Keep Their Perimeter Running Around the Park: Introducing Perimeter	Find the perimeter of polygons (triangle, quadrilateral, pentagon, hexagon, etc.) given the side lengths; find an unknown side length of a regular polygon
Grade 4	
Balloon Flight: Understanding Decimal Numbers Zooming Decimals: Precision and Place Value	Read, write, order, and compare decimals to the hundredths place (0.15, 0.25, etc.)
Identity Properties: Exploring 0 and 1 Number Codes: Properties of Addition and Multiplication	Understand, use, and explain the commutative property ($9 \times 4 = 4 \times 9$), the associative property ($(2 \times 9) \times 5 = 2 \times (9 \times 5)$), and the identity property ($18 \times 1 = 18$) of multiplication
Jump Along: Multiplication on the Number Line Sum and Product Puzzles: Number Sense and Mental Computation	Understand, use, and explain the commutative property ($9 \times 4 = 4 \times 9$), the associative property ($(2 \times 9) \times 5 = 2 \times (9 \times 5)$), and the identity property ($18 \times 1 = 18$) of multiplication Know multiplication and related division facts through 12×12
Comparing Fractions: Number Sense and Benchmarks Fraction Tiles: Equivalent Fractions Jump Along: Equivalent Fractions on the Number Line	Compare, create, and explain equivalent fractions using manipulatives, models or illustrations (1/2 = 2/4 = 3/6 = 4/8 = 5/10, etc.) Compose (put together) or decompose (break apart) a given fraction into partial fractions with a common denominator using a variety of manipulatives, physical objects, or visual models (1/7 + 1/7 + 1/7 = 3/7 or 2/7 + 1/7 = 3/7 and 3/8 = 1/8 + 1/8 + 1/8 or 3/8 = 2/8 + 1/8)
Angle Measurement: Estimation Practice	Draw points, lines, line segments, rays, angles (right, acute, obtuse), and perpendicular and parallel lines; identify these in two-dimensional figures
Angle Measurement: Introducing Protractors Measure by Degrees: Types of Angles	Draw points, lines, line segments, rays, angles (right, acute, obtuse), and perpendicular and parallel lines; identify these in two-dimensional figures Use a protractor to identify, classify, and draw right angles (90°), acute angles (less than 90°), obtuse angles (greater than 90°), straight angles (180°), and perpendicular and parallel lines

New York City Title I Elemetary School Activities
© 2012 Key Curriculum Press

Activity Title	Learning Objective
Grade 4, *continued*	
Straight Ahead: Segments, Lines, and Rays Mondrian in Motion: Parallel and Perpendicular Lines	Draw points, lines, line segments, rays, angles, and perpendicular and parallel lines; identify these in two-dimensional figures
Perimeter Formulas: Algebraic Notation Pool Border: Equivalent Expressions	Find the perimeter and area of polygons using mathematical formulas; Use equations involving a variable (x) to solve for an unknown side
Grade 5	
How Close Can You Get: Rounding Decimals	Identify the place value for each digit in decimals to thousandths; use this understanding to estimate and round decimals to the nearest tenth ($0.080 \rightarrow 0.1$), hundredth ($0.136 \rightarrow 0.14$), and thousandth ($0.1257 \rightarrow 0.126$)
Combination Locks: Factors, Composites, and Primes How Many Bugs: Divisibility and Remainders	Identify the factors of a given number (12: 1, 2, 3, 4, 6, 12)
Mystery Number: Multiples and Factors	Identify the factors of a given number (12: 1, 2, 3, 4, 6, 12) Use a variety of strategies to multiply multi-digit whole numbers using standard algorithms (step-by-step instructions for computing) to solve problems and check results; illustrate and explain the calculations by using equations, rectangular arrays, and/or area models
Two-Digit Multipliers: Visualizing the Distributive Property	Use a variety of strategies to multiply multi-digit whole numbers using standard algorithms (step-by-step instructions for computing) to solve problems and check results; illustrate and explain the calculations by using equations, rectangular arrays, and/or area models
Magic Multiplying Machine: Exploring Multiplication Magic Dividing Machine: Exploring Division	Use a variety of strategies to add, subtract, multiply, divide decimals to thousandths
Jeff's Garden: Area Model of Fraction Multiplication	Use a variety of strategies to multiply fractions ($3/4 \times 1/2 = 3/8$)
Function Machines: Introducing Functions Function Machines: Working Backward	Identify, describe, extend, and create numeric patterns using two rules, functions, and geometric patterns; analyze the pattern or whole-number function and state the rule given on a table or an input/output box
Perfect Packages: Surface Area and Volume Prism Dissection: Surface Area Pyramid Dissection: Surface Area Stack It Up: Volume of Rectangular Prisms	Compose (put together) and decompose (take apart) three-dimensional figures to explore formulas for volume ($V = L \times W \times H$) and surface area (surface area = sum of areas of surfaces)
Lulu: Introducing the Coordinate Grid Coordinate Patterns: Points on a Line	Know how to write and plot ordered pairs (x, y) correctly in the first quadrant of a coordinate grid
Finish the Polygon: Concept of Area	Find the perimeter and area of regular and irregular polygons using mathematical formulas; use equations involving an unknown variable (x) to solve for an unknown side

Activity Title	Learning Objective
Grade 5, *continued*	
Angle Measurement: Estimation Practice	Measure and draw angles using a protractor; find the measure of a missing angle
Point Graphs: Representing Data	Represent and interpret data sets on a line plot, line graph, coordinate graph, and bar graphs: Construct line graphs from data sets; Include axis, labels, and scales
Target Mean Game: Data Distribution and The Mean Mean Meets the Median: Measures of Central Tendency	Represent and interpret data sets on a line plot, line graph, coordinate graph, and bar graphs: Construct line graphs from data sets; Include axis, labels, and scales Calculate, interpret, and solve problems using the mean and mode for a given set of data

Grade 1 Activities

Skip-Counting: Patterns on the Hundred Chart

INTRODUCE

Project the sketch on a large-screen display for viewing by the class. Expect to spend about 15 minutes.

1. Open **Skip-Count Hundred Chart.gsp.** Go to page "Patterns" to introduce the model. If students are not familiar with the hundred chart, ask them to describe what they see. [The numbers 1 through 100 displayed in rows of ten numbers] Accept all observations and note the language that students use.

2. Ask questions to spark students' interest. As you take responses, don't give away what the machine does. Explain, *This is a skip-counting machine. It has a* **Stop** *and* **Start** *button. What do you think the machine might do if we press the button?* Students may guess that the machine will skip-count and "show the numbers" as it counts.

   Suppose it does skip-count? Can we know what number it will count by? Students may guess that the *Count by* = 5 parameter they see on screen tells the number the machine will count by. *If the machine does count by 5's, what do you think the chart will look like when the machine counts?* Students who are comfortable with counting by 5's and who are familiar with the hundred chart will be quick to point out that all of the count-by numbers for 5 on the chart are in the fifth and tenth columns.

   Shall we find out? Are you ready? Press *Start* and *Stop* several times, giving students time to observe that the multiples of 5 are highlighted, one at a time, in order. Stop before the count reaches 75 and ask, *When I press the button again, what other numbers will the machine highlight?* Take responses and then let the machine count to 100. *What do you notice?* Elicit these two observations.

   • Counting by 5's highlights two columns, one in the middle and one at the right side (or the fifth and tenth columns).

   • The ones-place digits in the multiples alternate between 5 and 0.

Introduce the terms ones-place digit and tens-place digit, if necessary.

3. Most important is that students explore why counting by 5's produces these patterns. Students may volunteer explanations without being prompted. Here is a sample.

   The 5's are in two columns. The middle column has the numbers that have 5 in the ones place. All those numbers are an odd number of 5's: one 5, three 5's, five 5's, to the end of the chart. The last column has the numbers that have 0 in the ones place. That's because 5 and 5 makes 10, and two more 5's makes 10 more. And it goes down the chart like that.

If needed, ask questions like these to prompt students to think about patterns.

Why does counting by 5's make these two columns in the hundred chart?

Why does counting by 5's give numbers that have 5 or 0 in the ones place?

Does it make sense that there are twenty numbers highlighted when we skip-count by 5's to 100? Why?

Skip-Counting by 3

You may invite a student to the computer to "run" the machine.

4. Press *Reset* to clear the grid. Change the count by number to 3 by double-clicking the *Count By* parameter, changing the value to 3 in the dialog box that appears, and clicking OK. *Are you ready?* Press *Start and Stop*. When the count reaches any number close to 24, press the button again to stop the count.

If appropriate for your students, you may want to speed up the skip-count. Choose **Display | Increase All Speeds** one or more times after the count has started.

   Can you predict the next numbers the machine will show? Do you think you see any patterns? Talk with your neighbor about this for a minute, and then we'll start the machine again. When students are ready, press *Start and Stop* again to resume the count. Press again to stop at a number close to 60. Invite discussion. *Do you want to revise your predictions after seeing more of the skip-count numbers?* Resume counting, which will stop at 100.

5. *How would you describe what you see?* Again, students are likely to focus on the visual pattern first. Prompt them to relate the visual pattern to a numerical pattern. *Sam says the pattern looks like staircases. What did it look like to everyone else? I wonder why counting by 3's makes a staircase pattern in the hundred chart.* This is not as easy to figure out as the visual patterns created by the 2's, 5's, or 10's. For now, you are just modeling the behavior of wondering *why*; explain that students can think about this more when they work with the sketch themselves.

   What do you notice about the count-by-3's numbers? Here are two sample responses.

   The numbers switch back and forth between odd and even. That's because two odds make an even. And an odd and an even make an odd. So, four 3's is even—it's 3 plus 3 twice—and add another odd makes odd, 15.

The ones-place numbers repeat. It goes 3, 6, 9, 2, 5, 8, 1, 4, 7, 0 The first time it gets to 0 is at 30. Then the ones digits go up from 0 by 3's again, just like when the machine started.

6. Prepare students to work with the machine on their own. **We've started to explore patterns that show up when the machine counts by 3 and when it counts by 5. You're going to tell the machine numbers to count by and keep a record of your investigation. Choose numbers from 1 through 10 in any order you want. The machine likes whole numbers only!**

   Distribute the worksheet and explain how students will use it as they work with the machine. Read the sample entry in the table together. Point out that it records the count-by number and the patterns the students saw. Students should record the numbers they count by and the patterns they find.

   If you plan to have students print the patterns they generate, model the steps they will follow so that a chart prints on one page. With the 5's or 3's count showing on the hundred chart, choose **File | Print Preview**, select **Scale To Fit Page** (Mac) or **Fit To Page** (Windows), and click **Print.** Record these steps and post them where students can refer to them as they work.

DEVELOP

Expect students at computers to spend about 30 minutes.

7. Assign students to computers and tell them where to locate **Skip-Count Hundred Chart.gsp.**

8. Let students work at their own pace. If students need more room to record the patterns they find, have them write on the back of the worksheet or use another sheet of paper. As you circulate, here are some things to note.

   • Remind students, as needed, to choose whole numbers to explore. The machine is programmed to handle whole numbers only.

   • Pose questions such as these to learn about students' thinking and to encourage students to extend their thinking.

   What number are you trying next? How did you decide on that number?

   Is there a number it would be interesting to explore now that you've skip-counted by this number?

What do you predict? Why do you think that?

Tell me about the patterns you've found.

You've found two numbers that have a diagonal pattern. Do you have any ideas about why the patterns turn out to be diagonal?

What can you say about the ones digits in these skip-count numbers? Is there a pattern?

Student pairs may not have time to explore every skip-count number from 1 through 10. That's fine. The class will pool their findings in the class discussion.

- Take note of the ways students are exploring, predicting, conjecturing, and making connections. For example, some students may observe that some visual patterns (for example, stripes or diagonals) appear in more than one skip-count sequence; some students may make conjectures about why some sequences create the visual patterns they do; and some may have reasons for the numbers they choose to test. (*Let's try 4's and see whether the pattern is like the 2's, because two 2's makes 4.*)

- Note whether students are identifying numerical patterns and relationships, or visual patterns only. If needed, ask questions to direct students' attention to exploring patterns in the digits of a skip-count sequence on the chart and to the way a visual pattern relates to a numerical pattern.

SUMMARIZE

Project the sketch on a large-screen display for viewing by the class. Expect to spend about 30 minutes.

9. Students should have their worksheets with them. Open **Skip-Count Hundred Chart.gsp** and go to page "No Wait." On this page, when you enter a count-by number, all the multiples are highlighted at once, facilitating the class's discussion.

10. ***Share something with us about a skip-count pattern you found.***
Provide ample time for students to discuss their findings. Let students use the hundred chart to help them communicate to the class. Record students' ideas on chart paper. A few sample student descriptions and explanations follow. Expect that students will have identified many patterns they want to share.

This is a good time to introduce the term *multiple* if students have not used it before.

- *The 2's, 5's, and 10's made stripes. That's interesting, because you can make 10 evenly with 2's or 5's or 10's. Five 2's get you to 10. And two 5's get you to 10. So, then the machine is always starting over on the next row. The numbers don't wrap around the rows. None of the other* skip-count numbers up to 10 do that. They all go past 10 instead of landing on it. You can't make 10 out of them evenly.

- The 3's and 9's made staircases. We know that 9 is three 3's, so we thought maybe numbers you can make from 3 would make the same skip-count pattern. So, we tried 6. But it didn't turn out the same. We wondered why 6 doesn't make the same pattern. We think maybe it's because 6 is an even number, but 3 and 9 are odd numbers.

- In the 4's, 6's, and 8's, the numbers are like a knight's move in chess: two over and one down.

- When you skip-count by an even number, there are no highlighted numbers in the first column, the third column, the fifth column, the seventh column, or the ninth column. That's because all these columns have all the odd numbers. And you can't get an odd number when you add an even number to an even number.

- The ones-place digits for the 9's go down by 1: 9, 8, 7, 6, 5, 4, 3, 2, 1, 0. First, we didn't know why. Then we thought about counting by 10's. When you count by 9's, that's one less each time than when you count by 10's. So, first the ones-place digit is 1 less than 10 (9); then it's 2 less than 10 (instead of 20, you have 18); then its 3 less than 10 (instead of 30 you have 27); and we think that's why there's this pattern in the 9's.

11. Direct students' attention to the hundreds chart at the end of the worksheet. ***How long would it take you to shade in the skip-count numbers for counting by 4? What would make it easy? Hard? Are there any things you know right away about numbers on the chart you will fill in?*** Give students a moment to consider this without sharing out loud, and then explain that you'd like them to write their explanations so you can know what they are thinking. Distribute blank paper now, or have students write at a later time.

EXTEND

You may also suggest some challenges like the ones listed here.

The work in this lesson may have prompted new queries by students. Have them investigate these questions. Here are some examples.

Are there other numbers that you want to skip-count by? What interests you about those numbers?

Are there any numbers that aren't in the patterns so far? What are they? Why don't they show up?

Which count-by numbers land on 100? Which don't? Why?

Which number shows up in the most patterns? Which shows up in the fewest? Why?

Skip-Count on the
Hundred Chart

For
GSP5

Name:

Explore skip-counting on the hundred chart.

1. Open **Skip-Count Hundred Chart.gsp.** Go to page "Patterns."

2. Now you will tell the machine what number to count by.
   Double-click the count-by number on the screen.
   Enter a whole number in the dialog box and click OK.

3. Record any patterns you find. Use your own paper for more
   patterns. Here are two examples.
   3   diagonals
   3   The numbers switch between even and odd.

_____

_____

_____

_____

_____

_____

_____

_____

_____

_____

_____

_____

_____

_____

_____

_____

HUNDRED CHART

1	2	3	4	5	6	7	8	9	10
11	12	13	14	15	16	17	18	19	20
21	22	23	24	25	26	27	28	29	30
31	32	33	34	35	36	37	38	39	40
41	42	43	44	45	46	47	48	49	50
51	52	53	54	55	56	57	58	59	60
61	62	63	64	65	66	67	68	69	70
71	72	73	74	75	76	77	78	79	80
81	82	83	84	85	86	87	88	89	90
91	92	93	94	95	96	97	98	99	100

New York City Title I Elementary School Activities with The Geometer's Sketchpad
© 2012 Key Curriculum Press

Place-Value Counter:
Ones, Tens, Hundreds, Thousands

For GSP5   ACTIVITY NOTES

INTRODUCE

Project the sketch for viewing by the class. Expect to spend about 10 minutes.

1. Open **Place Value Counter 1000.gsp.** Go to page "Counter." Tell students, *Today you're going to use this special counter.* Press the + button in the ones place four times, pausing between presses. *What happens each time I press this button?* Students will observe that the number displayed by the counter increases by one.

2. *Watch what happens in the table.* Press the + button in the ones place four more times, again pausing between presses. Elicit the idea that the table is keeping track of the number of ones.

3. *I'll press another button.* Press the + button in the tens place several times. Students should observe that the number displayed increases by 10 each time this button is pressed, and the tens column of the table keeps track of the number of tens added to the number displayed.

4. Ask students to predict what will happen when you press the remaining two + buttons. Press each button several times to test students' ideas. The table keeps a tally of the number of hundreds or thousands added to the number displayed by the counter.

5. *I'm going to call the button on the right the ones-place button. Every time I press it, it increases the number shown on the counter by 1. What should we call the other buttons?* [the tens-place button, the hundreds-place button, and the thousands-place button]

6. Distribute the worksheet. Read the directions with the class. Model pressing *Reset* to clear the counter. Look at step 1 together. *Your goal is to make the counter display the number 10. When you've found one way, you'll try to find a second way.* Look at worksheet steps 2 and 3 together so students know what they are asked to do.

DEVELOP

Expect students at computers to spend about 35 minutes.

7. Assign students to computers and tell them where to locate **Place Value Counter 1000.gsp.** Tell students to work through steps 1−3 and do the Explore More task if they have time. Encourage students to ask a neighbor for help using Sketchpad if needed.

8. Let pairs work at their own pace. As you circulate, observe and listen to students' conversations. Here are some things to notice.

   - Some students may benefit from having base-ten blocks available as they work.

   - In worksheet step 2, how do students reach 100? Many students will likely start with Hundreds = 1, Tens = 0, Ones = 0. Thinking of 100 as ten 10s will lead them to Hundreds = 0, Tens = 10, Ones = 0. For students who need some help generating more ways, propose, *Suppose you pressed the tens-place button nine times instead of ten times.* Give them time to set that up in the model, and then ask, *Is there something you could do next to get to 100?*

   - In finding ways to reach 100 or 1000, students might be tempted to press the + button in the ones place many, many times. For example, Thousands = 0, Hundreds = 0, Tens = 0, Ones = 1000 certainly does the trick. Remind students that the worksheet limits the number of times they can press any single button to 100.

SUMMARIZE

Project the sketch. Expect to spend about 15 minutes.

9. Gather the class. Students should have their worksheets with them. Ask volunteers to share the ways they used to show 10 on the counter.

10. Next, ask for ways students used to show 100. Record solutions in a table like the one on the worksheet.

11. Write " _____ = 100" on the board. *Let's look at one of the ways in the table. How can we write this information as a mathematical statement using an equal sign and other mathematical symbols?* As an example, for a solution of 7 tens and 30 ones, some possible statements students might create are these.

$$7 \times 10 + 30 \times 1 = 100$$

$$70 + 30 = 100$$

$$(7 \times 10) + (30 \times 1) = 100$$

12. Write " _____ = 1000" on the board and have students represent one or more ways in their tables as mathematical statements.

13. *All of you found different ways to reach 100. I want you to look at Way 1 on your table for 100. I don't know what [Beth] wrote, but I'm going to tell her how she can change just one number in her table so that the numbers in Way 1 now equal 1000 instead of 100.* Whisper to one of your students to add 9 to the number in the hundreds place. Let the student make the change and have her confirm to the class that the new value of the numbers is 1000. ***What could I have whispered?*** Two other possibilities are that you told the student to add 90 to the number in the tens place or 900 to the number in the ones place.

EXTEND

What other questions about using the counter occur to you? Here are sample student queries.

Instead of finding just four ways to reach 100 or 1000, could we find ten ways?

How many ways total are there of getting to 100? To 1000?

How can we reach 10,000?

How can we reach numbers that don't end in zero?

Are there patterns that allow us to predict how many different ways there are to get to a particular number?

Suppose we're allowed to press any button only an even number of times. How many different solutions can we find?

Suppose we're allowed to press any button only an odd number of times. How many different solutions can we find?

Suppose we also had buttons that allowed us to subtract 1, 10, 100, or 1000? How many more solutions could we find?

What if we had buttons for adding 5 instead of 1?

ANSWERS

1. There are two possible solutions.

	Thousands	Hundreds	Tens	Ones
Way 1	0	0	1	0
Way 2	0	0	0	10

2. Sample solutions. Other solutions are possible.

	Thousands	Hundreds	Tens	Ones
Way 1	0	1	0	0
Way 2	0	0	10	0
Way 3	0	0	3	70
Way 4	0	0	5	50

3. Sample solutions. Other solutions are possible.

	Thousands	Hundreds	Tens	Ones
Way 1	1	0	0	0
Way 2	0	10	0	0
Way 3	0	9	9	10
Way 4	0	5	45	50

4. Answers will vary.

New York City Title I Elementary School Activities with The Geometer's Sketchpad
© 2012 Key Curriculum Press

Place-Value Counter

Name:

Find ways to show 10, 100, and 1000 on the counter.

EXPLORE

Open **Place Value Counter 1000.gsp.** Go to page "Counter."

Find more than one way to make the counter show each number.

Don't press a button more than 100 times!

Always press *Reset* to begin.

1. Can you find two ways to show 10?

	Thousands	Hundreds	Tens	Ones
Way 1				
Way 2				

2. Can you find four ways to show 100?

	Thousands	Hundreds	Tens	Ones
Way 1				
Way 2				
Way 3				
Way 4				

3. Can you find four ways to show 1000?

	Thousands	Hundreds	Tens	Ones
Way 1				
Way 2				
Way 3				
Way 4				

EXPLORE MORE

4. Think of a number to show. It can be a number that does not have 0 in the ones place. What is it? _____

   Can you find four ways to show it?

	Thousands	Hundreds	Tens	Ones
Way 1				
Way 2				
Way 3				
Way 4				

New York City Title I Elementary School Activities with The Geometer's Sketchpad
© 2012 Key Curriculum Press

Comparing Heights: Logic and Deduction

INTRODUCE

Project the sketch for viewing by the class. Expect to spend about 10 minutes.

1. Open **Comparing Heights.gsp.** Go to page "Heights." Enlarge the document window so it fills most of the screen.

2. Explain, *Today, you're going to compare the heights of four students—Ann, Bill, Carlos, and Denise. I'll give you some information about their heights, and you'll use that information to draw some conclusions.*

3. On the board, write: "Denise is shorter than Ann." Point out the same statement on the Sketchpad model. Drag the statement across the vertical line. *We'll use the four segments labeled with the students' names as a way to keep track of how their heights compare. What do you notice about the segments in their starting positions?* [They're all the same height.]

The actual heights of the segments do not matter.

4. Ask a volunteer to drag one of the four segments to match the statement "Denise is shorter than Ann." The student may drag the endpoint of segment "Ann" up or drag the endpoint of segment "Denise" down. Either way, the red star will turn green. *The star is red when the heights of the segments don't match the statements you have dragged across the line. The star turns green when you adjust the segments to match the statements.*

5. Drag across the line a second statement, "Ann is shorter than Bill." *Why did the star turn red again?* [Ann's segment is not shorter than Bill's segment. The segments don't match the statement.] Ask a volunteer to adjust a segment to make the statement true. The model will look something like the picture here.

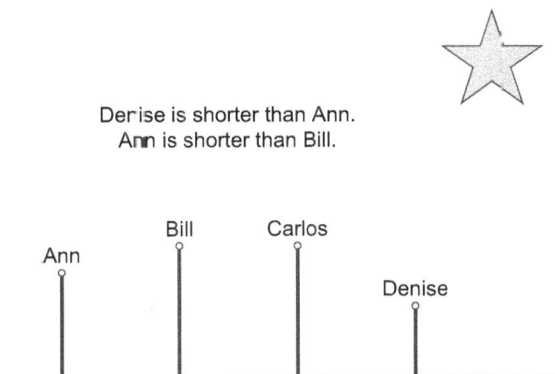

Denise is shorter than Ann.
Ann is shorter than Bill.

6. *I didn't give you a statement comparing Denise and Bill, but can you tell who is taller?* Take responses. By looking at the heights of the

segments, students will see that Bill is taller than Denise. Ask a volunteer to write a full sentence on the board comparing the heights of Denise and Bill.

7. Drag Ann's segment up and down. ***Why does the star sometimes turn red as I drag it?*** [If Ann's segment is shorter than Denise's segment or taller than Bill's segment, the star turns red.]

DEVELOP

Expect students at computers to spend about 35 minutes.

8. Assign students to computers and tell them where to locate **Comparing Heights.gsp.** Distribute the worksheet. Explain that in worksheet step 2 students should drag each statement in the first column of the table across the vertical divider line and then drag the segments to make the star turn green.

   If you have time, do step 4 and the Explore More. Encourage students to ask their neighbors for help if they have questions about Sketchpad.

9. Let pairs work at their own pace. As you circulate, here are some things to notice.

   • In problem 2a (in the table in worksheet step 2), students will drag the segments and learn that Ann is taller than Denise. Are they convinced that this will always be true, or do they adjust the segments to see whether another result is possible? A student who is convinced might say, *I know I can never make Ann shorter than Denise. If she's taller than Bill, and Denise is shorter than Bill, then the order of the heights must always be Ann–Bill–Denise.*

   • In problem 2c, do students realize that it is impossible to make any statement with certainty about the heights of Carlos and Ann? The star stays green regardless of whether Ann is taller, shorter, or the same height as Carlos, provided that Ann's segment is taller than Bill's segment.

   • Problem 2f is another example of a case in which it's not possible to draw a conclusive result based on the information given.

   • In problem 2g, the statements in the first column of the table are not identical to the wording on the Sketchpad model. Whereas the model says, "Ann is taller than Carlos," the worksheet says, "Carlos is shorter than Ann." This is purposeful. Observe whether students understand that the statements are equivalent.

• In worksheet step 3, students get practice working in reverse—starting with a visual representation and writing statements based upon it. There are many statements about the visual representation that students can make (12 in all).

SUMMARIZE

Expect to spend about 15 minutes.

10. Gather the class. Students should have paper and a pencil. Now students will solve similar logic problems without the Sketchpad model displayed. This provides an opportunity for students to transfer their experience of modeling and the logic involved to a setting away from the computer.

Write these two statements on the board.

• Bill is taller than Denise.

• Ann is shorter than Denise.

Ask students to represent the information in these statements on their papers. Students might draw vertical segments like the ones in the model, or they might devise another method. Ask, **Who is taller, Bill or Ann?**

Have students share their models. Was it necessary to write the name of each person alongside his or her segment? Could abbreviating the names (for example, A for Ann) be more efficient?

Continue to present problems of this type for students to solve. You can use some of the problems from worksheet step 2, make up your own problems, or share the problems that students made for each other in worksheet step 5.

11. **How did modeling help you to solve problems of this kind?** Close the activity with an opportunity for students to reflect on the usefulness of the visual representation presented and any models they devised themselves.

EXTEND

Page "Extend" connects logical reasoning with symbols and inequalities. The four segments on screen no longer represent student heights. The sentences from page "Heights" are replaced by inequalities with the letters A, B, C, and D.

Here are some suggestions for using the model.

- Ask students to read the inequalities aloud. *A is greater than B.*

- Ask students to "flip" several of the inequalities. A < B, for example, can also be written as B > A.

- Drag an inequality across the vertical line and ask students to drag a segment to make the star turn green.

- Have students use the model to reason about a set of inequalities. For example, if A > B and B > C, ask the students to write an inequality about A and C.

ANSWERS

2. a. Ann is taller than Denise.
   b. Carlos is taller than Denise.
   c. It is not possible to tell who is taller.
   d. Bill is taller than Carlos.
   e. Denise is taller than Ann.
   f. It is not possible to tell who is taller.
   g. Ann is taller than Denise.

3. Possible statements include these: Ann is shorter than Bill. Bill is taller than Carlos. Carlos is shorter than Denise. Ann is taller than Carlos. Ann is shorter than Denise. Bill is taller than Denise.

4. It is impossible for the statements "Ann is taller than Bill" and "Ann is shorter than Bill" to both be true. Other similar answers are possible.

5. Answers will vary.

6. If Bill is taller than Carlos and Carlos is taller than Denise, then Bill is taller than Denise. It is not possible for Bill to be shorter than Denise.

7. It is possible for up to six statements to be true at the same time. One possibility is the following: Ann is taller than Bill. Bill is taller than Carlos. Carlos is taller than Denise. Denise is shorter than Ann. Ann is taller than Carlos. Bill is taller than Denise.

Comparing Heights

For GSP5 Name:

Use a model and logic to solve some tricky problems!

EXPLORE

1. Open **Comparing Heights.gsp.** Go to page "Heights."

2. Read the "What You Know" statements in a row of the table below.
   In the model, drag the statements across the vertical line.
   Drag segments. Can you make all the statements in the row true?
   If you can, the star will turn green.
   Use the model to answer the question about who is taller.

What You Know	Who is Taller?	Answer
a. Ann is taller than Bill. Bill is taller than Denise.	Ann or Denise?	
b. Bill is shorter than Carlos. Bill is taller than Denise.	Carlos or Denise?	
c. Bill is shorter than Carlos. Ann is taller than Bill.	Carlos or Ann?	
d. Denise is taller than Ann. Ann is taller than Carlos. Bill is taller than Denise.	Bill or Carlos?	
e. Ann is shorter than Carlos. Bill is shorter than Denise. Bill is taller than Carlos.	Ann or Denise?	
f. Bill is shorter than Denise. Ann is taller than Carlos.	Ann or Bill?	
g. Carlos is shorter than Ann. Carlos is taller than Bill. Denise is shorter than Bill.	Ann or Denise?	

3. Here is the model showing how the heights of the four students compare. Write six true statements about the students' heights.

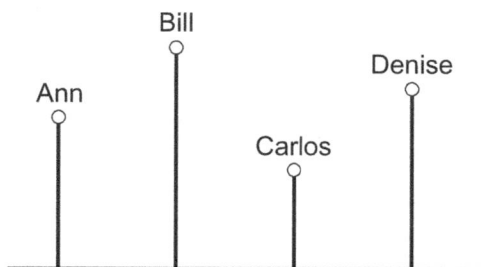

Bill

Denise

Ann

Carlos

_____

_____

_____

_____

_____

_____

4. Joan dragged two statements across the line. She found that it was impossible to make the star green. Which statements do you think she used?

_____

_____

EXPLORE MORE

5. Make up a problem like those in step 2 and have a classmate solve it.

_____

_____

_____

6. Can these three statements all be true? Explain. Why or why not?

- Bill is taller than Carlos.

- Carlos is taller than Denise.

- Bill is shorter than Denise.

_____

_____

_____

7. Drag as many statements as you can across the vertical line and make the star green. List the statements here.

_____

_____

_____

_____

_____

Grade 2 Activities

Circles and Squares: Representing an Unknown

INTRODUCE

Project the sketch for viewing by the class. Expect to spend about 15 minutes.

1. Open **Circles Squares An Unknown.gsp** and go to page "Introduction." Check that the values of the square and circle are 4 and 1, respectively. To change a value, double-click it, enter a new value in the dialog box that appears, and click OK. Ask students to describe what they see. The left side of the screen contains eight rows of squares and circles. A vertical line separates the objects from a column of zeroes. The bottom of the screen shows the numerical values of the square and circle.

   Explain, *We can drag any row of squares and circles that we want across the line. When I drag the first row, the computer will add the values of the square and circle.*

   Drag the square and circle across the divider line. The 0 will change to a 5, indicating the sum of 4 and 1.

2. Distribute paper. *I'd like each of you to record the information about the sum of the square and circle on your paper. There is more than one correct answer. You decide how to represent the information.* Ask volunteers to share what they wrote. Here are possible responses.

   5

   4 + 1 = 5

   ☐ ● 5

   ☐ ● = 5

   ☐ + ● = 5

3. If students do not offer the last statement above, ask them to write a statement that includes the square, the circle, an addition sign, and an equal sign. Read the statement together as a class: "Square plus circle equals 5." Drag the row back to the left of the line.

DEVELOP

Continue to project the sketch. Expect to spend about 30 minutes.

4. Focus on writing addition statements with symbols. Pick another row of circles and squares. Ask students to compute the sum of the symbols (before they are dragged across the vertical line) and write a corresponding statement. Ask volunteers to share their statements. Drag the row across the vertical line to check.

5. Write this statement on the board.

   □ + □ + ● = 9

It is important for students' future work in algebra that they develop the flexibility to see the same number sentence in different forms.

   Below it, write, "9 = ." Ask students to complete the sentence so that it gives the same information as the original statement. They might write,

   9 = □ + □ + ●

6. The order of the symbols can be changed without changing the result. Show the statement above and ask students to write another true statement using the same symbols in a different order. Here are two.

   9 = □ + ● + □

   9 = ● + □ + □

7. Return to the sketch. Pick other rows of symbols and ask students to write equivalent number statements. To add variety to the problems, press *New Values* to change the numerical values of the circle and the square.

Finding the Value of an Unknown Symbol

8. Go to page "Find the Unknown." Explain that the model is like the previous one, but now the value of the square is hidden. The goal is to figure out its value.

9. Drag the first row of symbols across the vertical line. For the sake of example, let's say that the value of the circle is shown as 7 and the reported sum is 10.

   Ask students to write the addition statement shown on their papers. Have a volunteer share the statement on the board.

   Is it possible for us to figure out the value of the square if we know the sum of the square and the circle? Students may suggest that one way to solve the problem is to replace the circle with its numerical value.

   □ + 7 = 10

   Students might solve this problem by phrasing it, for example, *What number plus 7 equals 10?* When the class has solved the problem, press *Show Answer* to reveal the value of the square.

10. Create new problems by pressing *New Challenge.*

© 2012 Key Curriculum Press

11. Increase the challenge by dragging either of the following rows. In both cases, the circles must be added in order to determine the value of the square.

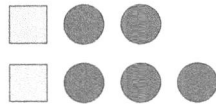

    ☐ ● ●
    ☐ ● ● ●

12. For an added level of difficulty, drag a row containing two or more squares. If, for example, ☐ ☐ ☐ ● ● = 23 and ● = 7, a sample solution path is to think, ☐ + ☐ + ☐ + 14 = 23. The sum of the three squares is equal to 9, so each square is equal to 3.

SUMMARIZE

Working away from computers, expect to spend about 15 minutes.

13. Have students write an explanation for someone who has never used the "circles and squares" model, explaining how to find the hidden value of the square. Students can include pictures and examples.

EXTEND

1. For students who would benefit from more individualized work, provide an opportunity to use the student sketch in pairs at a later time.

2. Pairs of students can create challenges for each other by changing the values of the symbols on page "Make Your Own." Now students can enter values for the square and circle that either exceed 8 or are negative; decimal values are rounded to the nearest whole number.

Bugs in Groups:
Dividing into Groups of Equal Size

INTRODUCE

Project the sketch for viewing by the class. Expect to spend about 10 minutes.

1. Open **Bugs in Groups.gsp,** and go to page "24 Bugs." Start with 5 as the value of the *Group Size* parameter.

   - Press *1 … 2 … 3 …* and let students observe the 24 bugs moving around.

   - Press *Red Light!* to make the bugs move into groups. Provide time for students to observe that the bugs arrange themselves in equal-sized groups except for four leftover bugs.

   - Press *1 … 2 … 3 …* to scatter the bugs.

   - Change the group size to 6 by selecting the *Group Size* parameter with the **Arrow** tool and pressing **+** on the keyboard one time.

   - Press *Red Light!* again. **What did you see?** Among other observations, students should note that the number of bugs in each full group is equal to the value of the *Group Size* parameter.

Students often like to try large values for the group size. The sketch accepts any value, including negative numbers, zero, and non-integer numbers. Because it doesn't make sense to have negative group sizes, or non-integer group sizes, encourage students to use natural numbers less than or equal to the number of bugs in all.

2. Ask, **Are there other group sizes that leave no bugs out of full groups?** Invite a volunteer to the computer to experiment with one or two other group sizes in response to suggestions from the class. Have the volunteer model pressing **+** or **−** on the keyboard one or more times to change the value of the *Group Size* parameter each time.

   Highlight the language that students use, and clarify vocabulary. The term *groups* can here refer to *equal-sized groups*, or *full groups*. The terms *remainder* and *left over* can be used to refer to any bugs that appear in a group that is not full. Experiment with enough values to elicit the observations that sometimes there are no bugs left over when equal groups are made and that the bugs always make as many equal-sized groups as they can.

3. Distribute the worksheet and read it through with the class. Let students know they can use more than one copy of the table if they need to. Make sure students understand that they should record every group size they try and the results. Go to page "Explore More" and point out that in worksheet step 5 students will use this page. Explain the differences in this model: Students can change the total number of bugs, and the buttons are different. Model pressing *Scatter* and then changing the total number of bugs by double-clicking *Bugs in All* and entering a new value in the dialog box that appears. Press *Group.*

DEVELOP

Expect students at computers to spend about 30 minutes.

4. Assign students to computers and tell them where to locate **Bugs in Groups.gsp.** Encourage students to ask a neighbor for help if they have questions about using Sketchpad.

5. Let students work at their own pace. Here are some things to be aware of as you observe and ask students to tell you about their strategies and thinking.

Notice that students generally do not employ repeated subtraction when solving division problems of this type.

- The problems students are solving are set in a division context and provide an opportunity for students to develop intuitive understandings about the relationship between multiplication and division. Students' own thinking, at this stage, is likely to involve multiplication strategies in which students "build up" to the number of bugs in all. In predicting or making sense of the number of groups, students may reason as in this example involving 24 bugs grouped by 6: *Six and six is 12, plus two more sixes is 12 more. That's 24.*

- Some students may try all possible group sizes, in order, starting from 1. Others may choose group sizes randomly. And others may be guided by their thinking about number composition and multiples.

- Students may intuitively construct the distributive property for multiplication over addition. *There are six groups of four. The three groups on my side are three times four—that's 12. The three groups on your side are also three times four. So, that's 12 plus 12—24 bugs.*

- Students may construct and apply the commutative property on their own. *We made 12 groups of 2 bugs. If we split the bugs in each group, we could make 2 groups of 12 bugs.*

- Students may notice and explain that they are using multiplication to figure out what will happen when they divide.

- Some students may overlook the possibility of a group size of one. Others may consider the possibility but debate whether it "counts" because "a group is more than one." Simply ask students to consider that a group size of one leaves no *remainder* bugs; each "group" is full—it has one bug.

- Students may also debate whether one group with all the bugs "counts" as grouping the bugs. This reasoning fits with students' experiences forming teams or dividing up objects into groups.

Again, simply ask students to entertain the criteria that no bugs are left out when the bugs form one group.

6. Students who have time should do the Explore More, which offers them the opportunity to investigate questions they pose. (Note that students often like to try large values for the *Bugs in All* parameter. The sketch accepts any value for the total number of bugs, but large numbers may make it hard to distinguish the bugs. Encourage students to use natural numbers of 100 or less.)

SUMMARIZE

Project the sketch. Expect to spend about 20 minutes.

7. Gather the class. Students should have their worksheets with them. Facilitate a discussion and have **Bugs in Groups.gsp** available for modeling. Here are ideas the discussion may bring out. You may wish to show a transparency of the table and record students' data for the class to refer to.

- For 24 bugs, students may say that there are no bugs left over when the group size is 2, 3, 4, 6, 8, or 12. There are also no left over, or *remainder*, bugs when the 24 bugs are grouped by 1's or by 24. (As mentioned in the Develop section, students may question whether a group of one is really a group, and whether the bugs are grouped when they are all in one group. Mathematically, 1 and 24 are factors of 24.)

- Students may invent and apply the commutative property of multiplication. *We saw that the numbers work in pairs. A size of one makes 24 groups, and a size of 24 makes one group; a size of two makes 12 groups, and a size of 12 makes two groups; there are eight groups of three, and three groups of eight; there are six groups of four, and four groups of six.*

- Students may be intrigued by the results of trying to group 11 bugs. *There were a lot of group sizes that had no bugs left out for 24, but only two sizes that worked for 11—1 and 11. One group with all the bugs and 11 groups that each had one bug. We wondered whether we could find other numbers like 11, so that's what we did when we used the "Explore More" page.*

- Students may have noticed that they were using multiplication to figure out and explain the results of dividing the bugs into groups.

Highlight this idea, inviting several students to express their thinking in their own words.

8. If students had time to pose their own questions and explore them, have them share their questions and any findings, as time permits.

EXTEND

1. *What other questions can you ask about dividing objects into groups of equal size?* Encourage student curiosity and provide time for students to explore their questions using the sketch. Here are some sample student queries.

   Is there a way to know whether there will be bugs left over before you group them?

   Why can the number of groups and the size of the groups be switched?

   For certain numbers of bugs like 11 or 13, we can only make equal groups with no bugs left out if we put the bugs all in one group, or if we make groups of one. Why is that?

   Do odd numbers of bugs work differently than even numbers of bugs?

   Can you know whether there will be an odd number of bugs or an even number left over?

   If I list all the group sizes that leave no bugs out for 16 bugs, is there an easy way for me to make a list for 32 bugs?

   What would happen if we made the group size number in the sketch larger than the total number of bugs?

2. Have students go to page "26 Bugs." Pose this question: *Now there are 26 bugs. Can you find a value for the group size that leaves exactly two remainder bugs? Can you find other values that leave exactly two remainder bugs?* Encourage students to see whether there is a pattern to the possible values. [Group sizes of 4, 6, 8, 12, and 24 all leave two remainders. These are numbers that are factors of $26 - 2 = 24$ (but aren't also factors of 26).] *Pose another problem: Can you find values that leave exactly three remainder bugs?* Group sizes that have three remainders will be numbers that are factors of $26 - 3 = 23$; since 23 is a prime number, only a group size of 23 will have three remainders.)

3. Have students who would benefit from additional work dividing bugs into groups of equal size use the sketch, including the additional pages that display 23, 26, and 30 bugs.

ANSWERS

2. 1, 2, 3, 4, 6, 8, 12, and 24

4. 1 and 11

5. Answers will vary.

Bugs in Groups

Name:

What happens when the bugs try to make groups with no bugs left out?

EXPLORE

Use the table on the next page to record the ways you group the bugs.

1. Open **Bugs in Groups.gsp.** Go to page "24 Bugs."

2. What group sizes leave no bugs out of full groups?

   _____

3. Go to page "11 Bugs."

4. What group sizes leave no bugs out of full groups?

   _____

EXPLORE MORE

5. Go to page "Explore More." Make the number of bugs any number less than 101.

   Think of a question to explore. Write it here.

   _____

   _____

   Explore your question. Write about what you find out.

   _____

   _____

   _____

   _____

For GSP5

Bugs in Groups

Bugs in All	Number of Groups	Group Size	Left Over

New York City Title I Elementary School Activities with The Geometer's Sketchpad

How Much Is Half: Size of the Unit Whole

ACTIVITY NOTES

INTRODUCE

Project the sketch for viewing by the class. Expect to spend 10 minutes. If students are not familiar with the circle model of fractions, spend additional time exploring this representation.

1. Open **How Much Is Half.gsp.** Go to page "One Circle." Make sure the fraction circle represents $\frac{1}{2}$. Familiarize students with the model, encouraging discussion and eliciting students' prior understanding of the meaning of the numerator and denominator of a fraction. Here is one way to introduce the model.

   • Select the denominator. Press the + key on your keyboard once. The fraction is now $\frac{1}{3}$.

   • Continue to press the + key on your keyboard, pausing each time, until the circle shows eighths. Listen for students' comments that as the denominator gets larger, the parts get smaller.

   • Students may propose trying some large numbers for the denominator. Press the + key repeatedly and go all the way to 50ths.

   • Double-click the denominator and, in the dialog box that appears, enter the value 100 and click **OK.** Students enjoy seeing that a value of 100 makes it almost impossible to see the parts.

   • Double-click the denominator and change its value to 8.

   • Ask what students think the fraction circle will look like if you change the numerator to 2. Take responses. Select the numerator, press the + key once, and discuss the results.

Remember that the model does not work for fractions greater than 1.

   • Continue to press the + key on your keyboard, pausing each time. Listen for students to comment that as the numerator gets larger, more parts are colored. Also, because the denominator has not changed, the size of the parts remains the same.

   • Show the fraction circle when both the numerator and denominator are 8. Elicit the idea that $\frac{8}{8}$ is the same as 1 whole.

   • Introduce the slider, focusing on the behavior of the model. Students should observe that dragging the slider's endpoint left and right changes the size of the circle, but not the number of parts colored. Explain that the slider controls the distance from the center of the circle to its border; changing that distance changes the size of the circle.

New York City Title I Elementary School Activities with The Geometer's Sketchpad
© 2012 Key Curriculum Press

37

DEVELOP

Expect students at computers to spend about 20 minutes.

2. Pose this problem: *Malik has $\frac{1}{2}$ of a pie and Laura has $\frac{1}{2}$ of a different pie. Laura's slice is larger than Malik's slice. How can that be?*

3. Go to page "Two Circles." *Let's model this problem.* Have students talk in pairs about how they would model it. Invite a volunteer to the computer to try.

   Each circle should show $\frac{1}{2}$, and the circles should be different sizes. Students can drag one circle onto the other to directly compare the halves. Elicit responses. *Who wants to say in their own words what was modeled? What does this demonstration show?* Here are samples of student thinking.

   You don't know what size $\frac{1}{2}$ is until you know what it is half of. You have to know what the whole thing is.

   A half can be different sizes. It depends on how much it's half of.

   It's like walking halfway home or walking halfway across the United States!

Elicit the idea that there are many ways the model could show one-fourth of one circle being larger than one-half of the other.

4. Press *Reset* so both circles are again the same size. *I have $\frac{1}{4}$ of a pie. My friend has $\frac{1}{2}$ of different pie. Is it possible that I have more pie?* Ask students to visualize what the model might look like when it answers this question, and then have them share their thinking in pairs. Invite a volunteer to the computer to demonstrate.

5. *What are some other questions about the size of fractions that you could model with these fraction circles?* Tell students that the questions may be about real-life situations but don't have to be. Have students work for a few minutes in pairs or small groups.

6. Ask the pairs or groups to offer questions for the class to explore. Have a volunteer or volunteers use the model to explore each question. When an appropriate demonstration has been given, take the time to have students restate what has been shown so all students have enough time to consider the ideas. Here are some of the types of questions the class might pose.

   • Malik has $\frac{1}{3}$ of a pie and Laura has $\frac{1}{4}$ of a different pie. Each of them says his or her own piece of pie is bigger. Laura is right. How can that be?

- Kyle eats $\frac{1}{8}$ of a cake. Sean eats $\frac{1}{2}$ of a cake. Kyle says that he's eaten more cake, but Sean doesn't agree. "If I ate half, then I ate more, obviously." Students will enjoy modeling cases in which a relatively small fraction needs to come from a much larger whole to be the greater quantity.

- Given an easy-to-use virtual model, students often propose extreme cases for consideration. They may pose a question like this one: *If we want to show that $\frac{1}{50}$ can be larger than $\frac{1}{2}$, I wonder how big the $\frac{1}{50}$ circle needs to be compared to the $\frac{1}{2}$ circle?* (Exploring an extreme case is worthwhile if the result is an engaging and memorable demonstration that the size of a fraction is related to the size of the whole.)

SUMMARIZE

Working away from computers, expect to spend about 15 minutes.

7. To help students consolidate their understanding, pose the following questions, recording the information for students to see. Provide thinking time.

   - *Sally has read $\frac{1}{3}$ of her book. Ben has read $\frac{1}{2}$ of his book. Who has read more pages?*

   - *One school bus is $\frac{1}{5}$ full. Another school bus is $\frac{1}{4}$ full. Which bus has more students on board?*

8. Facilitate discussion. Students should explain that it is impossible to answer the questions without more information. A student might explain, *We know the fraction, but we don't know the whole. We don't know how many pages are in each book or how many seats are in each bus.*

9. Have students create and share other questions of this type, situated in different contexts. You might have them write problems individually.

EXTEND

1. Present the first situation in step 7 above, revised. *Sally has read $\frac{1}{3}$ of a book that has 60 pages. Ben has read $\frac{1}{4}$ of a book that has 100 pages. Who has read more pages?* Students will recognize that this time the amount of the whole, the number of pages in each book, is given. Ask students to solve the problem in any way they like. When students find that Sally has read 20 pages and Ben has read 25 pages, ask them to rewrite the problem so that Sally has read more pages than Ben.

Present the school bus situation in step 7 again. Ask students to work with a partner and rewrite the problem to include a total number of seats in each bus. Present some of the problems for the class to solve.

2. Have the class consider situations in which the size of the whole is unknown and the size of the part is known. Ask students to explain what information is needed to answer each question. Expect students to need time and discussion to clarify what each problem is asking. In the first problem here, students will need to distinguish between the meanings of "a larger number of students" and "a larger fraction of the students."

   • In school elections, 100 students voted in one school, and 75 students voted in another school. In which school did a larger fraction of the students vote? [This cannot be answered. Students need to know the number of students in each school.]

   • Ramona has read 150 pages of her book, and Vanessa has read 200 pages of her book. Which girl has read more of her book? [This cannot be answered. Students need to know the number of pages in each book.]

Dynamic Triangles: Attributes of Triangles

For GSP5   ACTIVITY NOTES

INTRODUCE

Project the sketch for viewing by the class. Expect to spend about 15 minutes.

1. Open Sketchpad and enlarge a new sketch so it fills most of the screen. Explain, *Today you're going to use Sketchpad to explore triangles.* Follow these steps to construct a triangle in two different ways.

   • Using the **Point** tool, construct two points. *Are there enough points to draw a triangle?* [No] *How many points do we need?* [Three]

   • Construct a third point that could be a third vertex of a triangle. Using the **Text** tool, label the points *A, B,* and *C. How can I draw the triangle?* Invite a volunteer to construct the three sides of △*ABC* using the **Segment** tool.

   • Now construct a triangle another way: using only the **Segment** tool. Draw the first segment and label its endpoints *D* and *E*. Construct a second segment starting at point *D*, and label its other endpoint *F*. Finally, connect points *E* and *F*.

2. Using the **Arrow** tool, drag a side of the triangle and then a vertex. *Unlike a triangle we draw on paper, a triangle constructed with Sketchpad lets us change its size, shape, and orientation by dragging.*

3. *How would you describe a triangle?* Draw out the ideas that a triangle has three angles, it has three sides, and it is a polygon.

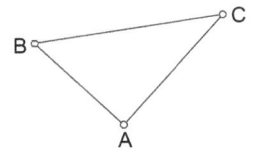

   Some students may believe that a triangle must have one horizontal side. Other students may think that a "skinny" triangle like the one shown at right isn't a triangle. Students will have the opportunity to expand their conceptions of triangles as they construct triangles and explore their behaviors.

4. Introduce the term *vertex* to describe the point where two sides meet. Say that *vertices* is the plural of vertex, so the triangle has three vertices. Introduce or review naming an angle by its vertex or by the three points that define the angle, with the second point always being the vertex. *You can name this angle "angle A," "angle BAC," or "angle CAB."*

5. *Do you see a right angle somewhere in this room?* Take responses. Students might point to the corner of a paper, a tile on the floor, or a doorframe. *What is a right angle?* Students may say that a right angle looks like a square corner or that it is an angle that measures 90°.

The words "looks like" are important here. Students aren't measuring the angles to prove that they are 90°.

Invite a volunteer to drag △ABC so that one angle looks like a right angle. Have another student do the same with △DEF.

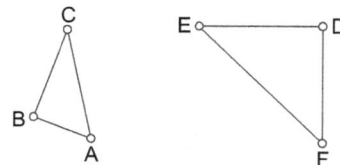

6. *In this activity you're going to make some special triangles. You won't take any measurements, so you'll be estimating when you want to make sides that are the same length or angles that are the same size. You'll use your drawings to make some educated guesses about triangles, like whether a triangle can have two sides that are the same length.*

7. If students will save their work, demonstrate choosing **File | Save As,** and let them know how to name and where to save their files.

DEVELOP

Expect students at computers to spend about 30 minutes.

8. Assign students to computers and tell them where to locate **Dynamic Triangles.gsp.** Students should have their worksheets. Tell students to work through step 4 and do the Explore More if they have time. Encourage students to ask their neighbors for help if they have questions about using Sketchpad.

9. Let pairs work at their own pace. As you circulate, here are some things to notice.

   • In worksheet step 2, students will construct a triangle using the **Segment** tool. With the first segment constructed, be sure students check that its intended endpoint is highlighted before clicking on it to attach the second segment.

   • In worksheet step 4, students move from page to page of the Sketchpad document and construct triangles based on the descriptions given. Remind students that they should not be concerned with the exactness of the side lengths or angle sizes. For now it is enough that a triangle looks about right and they think they can make a conjecture. *Without measuring, what do you think? Do you think a triangle can have three sides of different lengths?*

   • On each page, encourage students to make two different triangles that both fit the description. *How can you draw another triangle that looks like it has a right angle but has something different about its sides or angles from your first triangle?*

- Students may tend to make triangles that have one horizontal side, usually "the bottom." If you see them doing that, ask them to make a triangle with no sides that are horizontal.

- On page "One," ask students why the right angle in the triangle they made looks like a right angle.

- On page "Two," depending on how accurately students draw their triangles, they may or may not conjecture that a triangle can have two angles the same size. This idea will be revisited in the class discussion that follows.

- On page "Three," depending on how accurately students draw their triangles, they may or may not be able to conjecture that a triangle can have three angles the same size. This idea will be revisited in the class discussion that follows.

10. If students will save their work, remind them where to save it now. If a flash drive is available, save student sketches to share their triangles in the class discussion.

SUMMARIZE

Project the sketch and student sketches, if possible. Expect to spend about 15 minutes.

11. Gather the class. Students should have their worksheets with them. Start by giving students an opportunity to make some conjectures about triangles. Encourage students both to use what they have experienced in making triangles and to take some risks if they aren't sure about a conjecture. You might record these attributes on chart paper and put a check mark next to each attribute most of the class thinks is possible.

Do you think a triangle can have two sides of equal length?

Three sides of equal length?

No two sides that are the same length?

Do you think a triangle can have three angles that are the same size?

Two angles that are the same size?

Three angles that are all different sizes?

One right angle?

More than one right angle?

Two sides of equal length and two angles of equal size?

All sides of equal length and all angles of equal size?

Two sides of equal length and one right angle?

12. Open **Dynamic Triangles Present.gsp.** Go to page "One." *Does each of these triangles look like it has a right angle? Let's measure to see how close the drawing is.* (Students may have noticed that Sketchpad has a Measure menu.)

    Start with one of the triangles. Using the **Arrow** tool, select the three points that define the angle that looks like a right angle, selecting the vertex second. Choose **Measure | Angle.** Ask a volunteer to drag vertices of the triangle to try to come closer to or get exactly 90°. As the triangle is dragged, students will observe that the displayed angle measurement will update automatically.

13. Go to page "Two." *Let's take some measurements to check how close these triangles are. This time we need to measure the lengths of the sides.*

    To measure a segment, select it using the **Arrow** tool. (Don't select the endpoints.) Choose **Measure | Length.** If the measures of the two sides are not close, have a volunteer drag the triangle until the two measurements are closer or the same.

    What do you notice about the angles of these triangles? Take responses. Students will suggest measuring. Because the triangle will be close to isosceles, the measurements of the angles opposite the equal sides will be close to equal.

14. Go to page "Three." Measure the side lengths and have a volunteer drag the triangle so the three lengths are close in length or the same. *What do you notice about the angles of these triangles?* Students are likely to make the conjecture that the angles are the same size. Measure the three angles of one of the triangles. The angles will be close to or exactly 60°.

15. On pages "Four," "Five," and "Six," it is not necessary to measure angles or side lengths to be sure that the triangles fit the descriptions.

16. If time permits, discuss the Explore More. Have students share their designs, pointing out the different triangles they made.

EXTEND

What other questions can you ask about triangles? Encourage curiosity. Here are some sample student queries.

Can a triangle that has a right angle have two sides that are the same length? Three sides the same length?

What's the largest angle a triangle can have?

What's the smallest angle a triangle can have?

Do triangles that have two sides the same length always have two angles the same size?

If the sides of a triangle are all different lengths, are the angles all different sizes?

What is the sum of the angles of a triangle? Is it the same for every triangle or different for different triangles?

ANSWERS

Check student work. There are many ways students may draw triangles to satisfy each description.

Dynamic Triangles

For GSP5   Name: _____

See what kinds of triangles you can make.

EXPLORE

1. Open **Dynamic Triangles.gsp.** Go to page "One."

2. Construct a triangle.

   Label its three vertices.

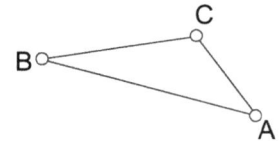

3. Drag the triangle by its sides and vertices. Make one angle look like a right angle.

   Draw another triangle that looks like it has one right angle.

4. The table shown here tells you what to make on the next pages.

   On each page, start by drawing a triangle and labeling its vertices.

   Drag the triangle to make the triangle you want.

   Then make another triangle that fits the description.

Page	Description
Two	Two sides look like they are the same length.
Three	All three sides look like they are the same length.
Four	One angle is larger than a right angle.
Five	All three angles are less than a right angle.
Six	All sides are a different length.

New York City Title I Elementary School Activities with The Geometer's Sketchpad
© 2012 Key Curriculum Press

EXPLORE MORE

5. Go to page "Explore More." Follow these steps to make and color the interior of a triangle.

    Draw a triangle.

    Select just the three vertices.

    Choose **Construct | Triangle Interior.**

    With the interior of the triangle selected, choose **Display | Color** and pick a color.

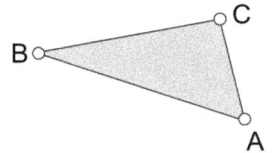

6. Add more triangles to your picture.

    You can change the color of segments by choosing **Display | Color.**

    You can change the thickness of segments by choosing **Display | Line Style.**

Dynamic Rectangles: Attributes of Rectangles

INTRODUCE

Project the sketch for viewing by the class. Expect to spend about 15 minutes.

1. Open **Dynamic Rectangles Present.gsp** and go to page "First Try." *Let's see whether we can construct a rectangle in Sketchpad using the* **Segment** *tool.* Refrain from describing any properties of rectangles as you quickly draw a quadrilateral that looks like a rectangle, using the **Segment** tool only. Hold down the Shift key as you draw the first three segments. Doing so will guarantee that the segments are horizontal and vertical. If your finished quadrilateral doesn't look quite like a rectangle, use the **Arrow** tool to adjust the vertices. Use the **Text** tool to label each vertex.

2. Get agreement from the class that the figure looks like a rectangle. Then, using the **Arrow** tool, drag each vertex of the figure. Students will observe that the figure can be dragged, resulting in many shapes that are not rectangles.

3. Go to page "Drag Test." Drag the vertices and sides of one quadrilateral and then of the other. Let students describe the results: The green quadrilateral remains a rectangle whereas the blue quadrilateral does not. *Both of these Sketchpad shapes start out looking like rectangles, but only one of them is the real thing. The one on the left is like the shape we drew: It can look like a rectangle, but it doesn't stay a rectangle when we drag its vertices. It fails the "drag test." What is different about the quadrilateral on the right?* Accept students' first thoughts. Here are samples of student thinking.

   It stays a rectangle when you drag. All the corners stay square corners.

   The sides opposite each other stay parallel.

   The sides opposite each other stay the same length as each other.

For your information, page "Pretenders 2" tells what type of quadrilateral each figure is. Expect debate about whether a square is a rectangle. *Can a rectangle have all sides of equal length?* [Yes] Some rectangles may be squares, and all squares are rectangles.

4. Go to page "Pretenders." *Here's a group of Sketchpad shapes. They all look like rectangles. But some of them are pretenders. Our job is to identify the pretenders.* Invite volunteers to the computer to perform the drag test on one figure at a time. As students identify which figures are pretenders (all but two are pretenders), their goal is to make two important observations: (1) In Sketchpad, shapes can behave differently (even if they look similar), and (2) observing the behavior of a shape when it is dragged reveals relationships between the sides and characteristics of the angles.

Accept students' informal language and introduce mathematical language as students describe why individual quadrilaterals are or are not rectangles. Let the discussion bring out these properties of rectangles.

- All angles are right angles. (A right angle looks like a square corner and measures 90°.)

- Adjacent sides (sides that meet at a vertex) are perpendicular. (Connect the term *perpendicular* to students' understanding of square corners, right angles, and 90°.)

- Opposite sides are of equal length.

- Opposite sides are parallel. (Connect the term *parallel* with the informal language students use, such as "always the same distance apart.")

To return to the original quadrilaterals, choose **Edit | Undo Animate Points.**

5. Go to page "Dancing." Press *Animate Quadrilaterals*. Again observe how students are thinking about the two quadrilaterals. Both quadrilaterals look like rectangles before they start to move, but only the red quadrilateral stays a rectangle when it starts to "dance." Explain what Sketchpad does when it animates. ***Animating in Sketchpad is like telling Sketchpad to drag all the vertices at once. The sides and vertices behave according to the relationships they have.*** Press the animation button again to stop the animation.

Review what students have observed about the shapes constructed using Sketchpad. ***In Sketchpad, shapes can behave differently. The behavior (what the shape does when it's dragged or animated, for example) reveals geometric properties of the shape. We might notice that opposite sides always seem parallel or that the angles always appear to be right angles.***

DEVELOP

Continue to project the sketch. Expect to spend about 20 minutes.

6. Go to page "Second Try." ***Our challenge is to construct a rectangle that will stay a rectangle when we perform the drag test or make it dance.*** Using the **Segment** tool, draw a horizontal segment. ***This is the first side of our rectangle.*** Using the **Text** tool, label the endpoints of the segment.

For your reference, pages "Construction 1" and "Construction 2" of **Dynamic Rectangles Present.gsp** outline other possible construction steps.

7. Follow students' ideas to construct each of the remaining sides of the rectangle. Having students lead the way engages them in using what they know about properties of rectangles as they learn features of Sketchpad. As the rectangle begins to take shape, keep asking, **What part of the rectangle do you want to make next?** As you take responses, look for one or two students to say what they expect or wish Sketchpad has a way to do. The following class discussion illustrates how a teacher facilitated as students tackled the construction.

Sample Class Discussion

Teacher: *We have one side of the rectangle, side AB. What do you want to make next?*

Student: *We need a side that starts at A and goes straight up.*

Student: *And makes a right angle with side AB.*

Teacher: *What word describes two lines that make a right angle?*

Student: *Perpendicular.*

Student: *Can Sketchpad make a perpendicular line?*

Teacher: *Let's look in the Construct menu.* (Chooses the Construct menu. All of the options are grayed out, but students are excited to see that **Perpendicular Line** is indeed one of the commands.) *A menu command is grayed out until Sketchpad has the information it needs to follow the command. To make a perpendicular line, Sketchpad needs some information. What do you think Sketchpad needs to know?*

Student: *It needs to know where the line should go.*

Student: *We want it go through point A.*

Teacher: (Selects point *A* and looks in the Construct menu. Students note that the **Perpendicular Line** command is still grayed out.) *This means that Sketchpad doesn't have all the information it needs. What else do we need to tell it?*

Student: *What the line is going to be perpendicular to—side AB.*

Teacher: (Rephrases the directions as she follows them) *I'm telling Sketchpad to construct a line through point A perpendicular to segment AB.* (Selects segment *AB*. Now both the segment and point *A* are selected. Chooses the Construct menu again.)

Students: *Now we can choose* **Perpendicular Line**! (Teacher chooses the command.)

Student: *Why doesn't the line have endpoints?*

Student: *Because a line goes forever in both directions.*

Teacher: *Let's see whether the line stays perpendicular to segment AB.* (Drags endpoint A and then drags the line that passes through A. The class observes that the line remains perpendicular to the segment.)

Student: *We need a segment for that side, not a line.*

Teacher: *That's right, and we'll see how to fix that a little later. Let's continue. What do you want to construct next for the rectangle?*

Student: *The side that starts at point B. So, a line going through point B.*

Student: *Not just any line. The line has to be perpendicular to side AB.*

Student: *Select point B and side AB and choose* **Construct | Perpendicular Line.**

Teacher: (Rephrases the directions as she follows them) *I'm telling Sketchpad to construct a line through point B perpendicular to segment AB.*

Student: *But, we could also make this side a different way, by making a line that's parallel to the line going through A.*

Student: *Right, because it says* **Parallel Line** *in the Construct menu, we know Sketchpad makes parallel lines.*

Student: *We could select point B and the line through point A and choose* **Construct | Parallel Line.**

Teacher: (Gives the class time to consider this option) *Let's try that. I'm going to choose* **Edit | Undo** *to back up. Now I'll make sure point B and the line through A are selected, and then I'll choose* **Construct | Parallel Line.** (Carries out those steps, and lets students confirm that this method also works)

Teacher: *Let's see whether the line stays perpendicular to segment AB.* (Drags endpoint A. The class observes that the line remains perpendicular to the segment.)

Teacher: *What do you want to construct next?*

Student: *The last side: The side parallel to side AB.*

Student: *Let's use the* **Parallel Line** *command again.*

Student: *But wait, don't we need a point? Can we use the tool that looks like a point to draw one?*

Teacher: **Let's place a point directly onto one of the lines.** (Selects the **Point** tool and moves it over the line through *B*. When the line is highlighted, she clicks to place the point and then labels the point as *C* with the **Text** tool.) **If I drag this point, it stays on the line.** (Selects the point with the **Arrow** tool and drags it)

Student: *Select point C and line BC and choose* **Construct | Perpendicular Line.**

Student: *Or we could select point C and segment AB and choose* **Construct | Parallel Line.**

Teacher: **Will both ways work? Let's hear them again.** (Students repeat the ways.) **Let's try both ways.** (Tries the first method. When the class verifies that this works, she chooses **Edit | Undo Perpendicular Line** and tries the second method, which also works.)

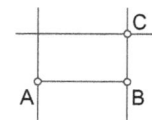

Teacher: **Now we need the last vertex.** (Using the **Arrow** tool, clicks on the intersection to construct point *D*; labels the point)

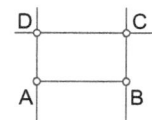

Student: *Three of the sides are lines. We need them to be segments.*

Teacher: **We can replace the lines with segments.** (Selects all three lines, but not the vertices or segment *AB*, and chooses **Display | Hide Lines;** then connects the points using the **Segment** tool)

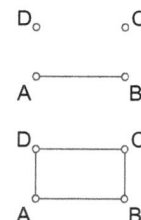

SUMMARIZE

Continue to project the sketch. Expect to spend about 10 minutes.

8. When your class has completed the rectangle construction, ask, *Are you ready to try the drag test?* Invite a volunteer to drag the figure's vertices. Assuming the construction is correct, students will observe that the rectangle stays a rectangle. *What stays the same?* [It stays a rectangle.] *What can change as the vertices are dragged?* [The size, shape, and orientation]

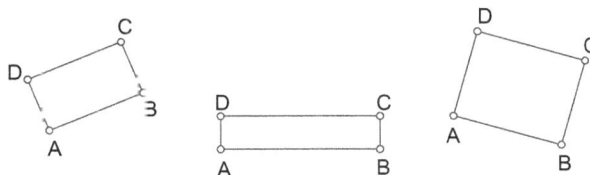

Now make the rectangle dance. With the **Arrow** tool, drag over the rectangle to select it, and then choose **Display | Animate.**

If the construction does not stay a rectangle, choose **Edit | Undo** repeatedly to back up and look for any errors. Then construct it correctly.

9. Go back to page "Drag Test." Drag the vertices of each figure so that students can identify which figure stays a rectangle and which does not. Facilitate discussion in which students reflect on what they have learned.

   Which quadrilateral is like the one we just constructed? [The green one] *Which is like the first quadrilateral we drew—using just the Segment tool?* [The blue one]

   Why do these quadrilaterals behave differently? Encourage many students to express in their own words what they understand about constructing a figure in Sketchpad. Elicit the idea that the sides and angles maintain the relationships Sketchpad used when it constructed them.

Students might like to imagine they are explaining the answers to these questions to a class visitor who is seeing Sketchpad for the first time.

   What can we find out by performing the drag test? Again, invite many students to respond. Elicit the idea that dragging the figure reveals the relationships between parts of the figure.

10. If time allows, go back to page "Pretenders." Choosing one quadrilateral at a time to investigate, drag the vertices and have students point out what is revealed. What relationships do the sides have? What is true of the angles? This exploration reinforces understanding of the nature of constructing in Sketchpad while previewing or developing understanding of properties of other quadrilaterals.

EXTEND

1. *What other questions about rectangles can you ask?* Encourage curiosity. Here are some sample student queries.

   Are there other shapes besides squares that are rectangles?

   Are there other ways to make a rectangle using Sketchpad?

   How do you construct those other shapes, like a parallelogram and a rhombus?

   Why is it true that when all the angles are right angles, the opposite sides are parallel and the same length?

2. Provide an opportunity for student pairs to construct a rectangle on their own. Encourage students to ask each other for help if they have questions about using Sketchpad or about the construction. Expect younger students to need more support; if parallel and perpendicular lines are new to them, post a picture to clarify.

3. Construct a square. This is a nice follow-up to the rectangle construction because it challenges students to use what they've learned when constructing a rectangle to build a related quadrilateral. Follow these steps to start the construction.

   • Open a new sketch.

   • Use the **Compass** tool to draw a circle.

   • Use the **Text** tool to label the center of the circle and the point on its circumference.

   • Use the **Segment** tool to connect points *A* and *B*.

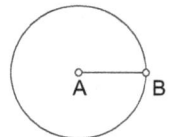

   Ask students how to continue the construction to build a square with side *AB*. Students will have to think about the role of the circle in creating segments of equal length. For your reference, a step-by-step construction is shown on page "Square Construction."

Square or Not: Properties of Squares

INTRODUCE

Project the sketch for viewing by the class. Expect to spend about 10 minutes.

1. Explain, *Today you're going to be exploring squares. What do you know about squares?* Here are some sample student responses.

   A square has four right angles.

   A square has four 90-degree angles.

   All four sides of a square have the same length.

   A square has four right angles and four equal sides.

2. Open **Square or Not Present.gsp.** Go to page "Sample." Ask students whether the quadrilateral looks like a square. Most students will probably agree that it does.

   This quadrilateral certainly looks like a square, but maybe it's just pretending to be a square. Let's check: One by one, I'm going to drag the four vertices of the quadrilateral. If ABCD is a "real" square, then it will stay a square no matter how I drag it. Let's try.

   Drag, one at a time, points *A*, *B*, *C*, and *D*, saving point *D* for last. Dragging points *A*, *B*, and *C* changes the size and orientation of the quadrilateral, but in all three cases, it still looks like a square. Only when point *D* is dragged does *ABCD* reveal itself to be a "pretend" square because angle *D* does not stay 90°.

   With students' help write the following statement on the board: "*ABCD* does not stay a square when point *D* is dragged." Explain that it will help students to refer to quadrilaterals and vertices by their names as they complete their worksheets.

3. Go to page "Test the Quadrilaterals."

   There are five quadrilaterals here. At least one of them is a 'real' square: No matter how you drag it, it will remain square. But some of these quadrilaterals are just pretending to be squares. When you drag them, they won't always stay square.

   Your job is to figure out which of these quadrilaterals are real squares and which are pretending to be squares. You'll need to drag all four vertices of each quadrilateral, one at a time.

4. If you want students to save their Sketchpad work, demonstrate choosing **File | Save As,** and let them know how to name and where to save their files.

DEVELOP

Expect students at computers to spend about 30 minutes.

5. Assign students to computers and tell them where to locate **Square or Not.gsp.** Distribute the worksheet. Tell students to work through steps 1–8. Encourage students to ask a neighbor for help if they have questions about using Sketchpad.

6. As you circulate, listen to students' conversations. Here are some things to notice.

Only quadrilateral *UVWX* is a "real" square.

- Do students recognize that the five quadrilaterals, in their initial state, look like squares? Some students may not think so because of the squares' orientations on the page. One of the primary goals of this activity is to help students develop the ability to recognize squares in all sizes and orientations.

- As students explore a quadrilateral, are they checking what happens when they drag each of the four vertices, one at a time? *It is important that students drag each vertex.* In quadrilateral *IJKL*, for example, students will only learn that it is a rectangle by dragging point *K*.

- What visual cues do students use to judge whether a quadrilateral is "really" a square? For students who are having trouble, ask, **If this quadrilateral is a real square, what has to be true about the lengths of its sides when you drag it? What has to be true about its angles?**

- Students may correctly note that it's impossible to tell whether a quadrilateral is a square without being able to measure its lengths and angles. Explain, **For now, base your conclusions on what you see. We'll take some measurements later.**

- Depending on students' experience with quadrilaterals, they may identify *IJKL* as a rectangle, *MNOP* as a parallelogram, and *QRST* as a rhombus. If not, don't worry; the main goal of this activity is to focus on squares.

- To return to the initial state of the sketch with all five quadrilaterals looking like squares, press *Reset.*

SUMMARIZE

Project the sketch. Expect to spend about 20 minutes.

7. Gather the class. Draw a table on the board like the one that follows.

Quadrilateral	Like a Square	Not Like a Square
EFGH		
IJKL		
MNOP		
QRST		
UVWX		

Let's use this table to collect information about the quadrilaterals. For each quadrilateral, let's record why it's like a square and why it's not like a square.

Start with quadrilateral *EFGH*. Ask volunteers for reasons why the quadrilateral is, or is not, like a square. As students answer, they can demonstrate with the Sketchpad model. Sample responses follow.

Even though it's just pretending, the quadrilateral EFGH does look like a square before you drag its vertices.

Quadrilateral EFGH is not like a square because it doesn't stay a square when you drag its vertices. It doesn't keep its right angles and its side lengths don't stay equal.

Continue filling in the rest of the table, having volunteers demonstrate with the Sketchpad model. Sample responses follow.

Quadrilateral	Like a Square	Not Like a Square
IJKL	Always has right angles	Sides are not always equal in length
MNOP	Opposite sides stay parallel	Doesn't always have right angles Sides are not always equal in length
QRST	Sides are always equal in length	Doesn't always have right angles
UVWX	Always has four right angles Always has equal side lengths	

8. Measure the side lengths and angles of square *UVWX*. To measure the side lengths, select all four segments with the **Arrow** tool (but not the vertices) and choose **Measure | Length.** Measure each of the four angles individually. To measure ∠ *U*, select in order points *V, U,* and *X,* and choose **Measure | Angle.**

   With all the side lengths and angle measures displayed, drag point *V.* As the square changes size, ask students what they notice. [All four side lengths change but remain equal to each other. All four angles stay 90°.]

9. If students are familiar with the terms *rectangle, parallelogram,* and *rhombus,* you can ask them to identify, by name, the blue, purple, and green quadrilaterals.

ANSWERS

2. Quadrilateral *EFGH* is not a real square. When its vertices are dragged, the angles do not remain 90° and its side lengths do not stay equal to each other.

3. Quadrilateral *IJKL* is not a real square. When vertex *K* is dragged, the angles remain 90°, but its side lengths do not stay equal. The quadrilateral is a rectangle.

4. Quadrilateral *MNOP* is not a real square. When vertex *M, O,* or *P* is dragged, its sides remain parallel, but its angles do not remain 90° and its side lengths do not stay equal. The quadrilateral is a parallelogram.

5. Quadrilateral *QRST* is not a real square. When vertex *Q, S,* or *T* is dragged, its side lengths stay equal to each other, but its angles do not remain 90°. The quadrilateral is a rhombus.

6. Quadrilateral *UVWX* is a real square. When its vertices are dragged, the angles remain 90°. The size of the quadrilateral grows and shrinks, but its side lengths always remain equal to each other.

Square or Not

Name:

Some quadrilaterals are pretending to be squares. Can you find the "real" squares?

EXPLORE

1. Open **Square or Not.gsp.** Go to page "Test the Quadrilaterals."

2. Drag vertex *E* of quadrilateral *EFGH.*
   Now drag another vertex. One by one, drag each vertex.
   Do you think it is a real square? Why or why not?

   _____

   _____

   _____

3. Check quadrilateral *IJKL,* dragging one vertex at a time.
   Do you think it is a real square? Why or why not?

   _____

   _____

   _____

4. Check quadrilateral *MNOP.*
   Do you think it is a real square? Why or why not?

   _____

   _____

   _____

5. Check quadrilateral *QRST.*
   Do you think it is a real square? Why or why not?

   _____

   _____

   _____

6. Check quadrilateral *UVWX*.

    Do you think it is a real square? Why or why not?

    _____

    _____

    _____

7. Color the inside of the real squares.

    For each real square, select its four vertices in either a clockwise or counterclockwise direction. Then, choose **Construct | Quadrilateral Interior.** You can change the color by choosing **Display | Color.**

8. Animate each quadrilateral to see how it behaves. Select the four vertices of a quadrilateral and choose **Display | Animate Points.**

    Use the arrows on the Motion Controller to change the speed of the animation. Press the Stop button to end the animation.

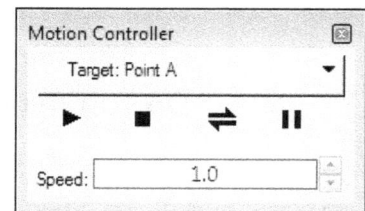

Motion Controller	☒
Target: Point A	▼
► ■ ⇌ ❚❚	
Speed:	1.0

Circles All Around: Parts of a Circle

INTRODUCE

Project the sketch for viewing by the class. Expect to spend about 15 minutes.

1. Open **Circles All Around Present.gsp.** Go to page "Compass Tool." Enlarge the document window so it fills most of the screen.

2. Use the **Compass** tool to construct a circle. *What is this figure?* [A circle] *Where can you find circles around you?* Students may suggest a clock, a plate, or a bike tire, for example.

The point on the circumference is sometimes called the *radius point* because it determines the radius of the circle.

3. Indicate the two points, one at the center of the circle and one on the circumference of the circle. *Let's see what happens when I drag each point.* Start by dragging the center point toward and away from the point on the circumference.

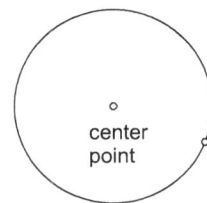

Now, drag the point on the circumference toward and away from the center point. *What stays the same and what changes as I drag the points?* Elicit mathematical language and students' prior knowledge of circles. Introduce and define terms, writing them on the board as they arise. Here are some sample student responses.

The circle always stays a circle, but it changes size.

There's a point on the border of the circle. The border of the circle is called the circumference.

To keep the circle the same size, drag its circumference.

The center point of the circle always stays at the center.

As the circle gets smaller, the circumference gets smaller.

As the circle gets smaller, the radius gets smaller.

4. Develop the concept that the *size* of a circle corresponds to the length of its radius. Use the **Segment** tool to construct a segment connecting the center of the circle to the point on the circumference. Emphasize that you need to check that each point is highlighted before clicking on it.

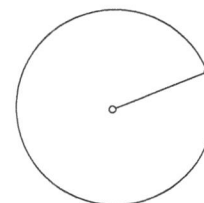

Now use the **Arrow** tool to drag the point on the circumference. *What happens to the segment when I drag this point away from the center?* [The segment gets longer.] *What happens to the size of the circle?* [It gets larger.] Repeat by dragging the point toward the center to illustrate that as the segment (the radius) gets shorter, the size of the circle gets smaller.

5. *Let's measure the length of the radius.* Select the segment (not the endpoints) and choose **Measure | Length.** The endpoints of the radius will be labeled *A* and *B*. Discuss how to read the measurement. The "m" stands for "measure."

   Now drag point *B*. *What happens to the measurement?* [It updates, increasing as the segment gets longer and decreasing as the segment gets shorter.]

6. Now construct another radius of the same circle. Using the **Segment** tool, start at the center of the circle and end anywhere on the circumference. (When the circumference is highlighted, the segment's endpoint will attach when you click.)

   Using the **Arrow** tool, drag the new point on the circle. *What happens when I drag this point?* [The point travels around the circumference of the circle, but the circle doesn't change size.]

7. Select the new segment and choose **Measure | Length.**

   What do you think will happen to the length of each radius when I drag point C? Take responses. Drag point *C* and have students note that because the circle stays the same size, both radii keep the same length. Then drag point *B* and note that the lengths of the radii change but always remain equal to each other.

   m $\overline{AB}$ = 5.15 cm
   m $\overline{AC}$ = 5.15 cm

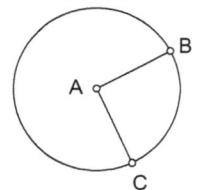

Remind students that the
plural of *radius* is *radii*.

8. *How many radii could we draw? Will they all be the same length?* Construct and measure a few more radii until the class understands that a circle has infinitely many radii and all are the same length.

   In mathematics we say that a circle is the set of all points in a plane that are the same distance from a center point. Explain that students can think of a plane as a flat surface. *Does this make sense?* Young students tend to think of a circle as a round object, but may not think of it in the way given by the definition. When asked to consider the definition, students may point out that all the radii have one endpoint on the circle, and those endpoints are the same distance from the center point: Their distance from the center is the length of the radii. You might ask, *Could we place a point on the sketch that isn't on the circle but whose distance from the center is the length of the radii?*

9. *Now you will construct some circles on your own and play two different circle games with a partner.* If you want students to save their work, demonstrate choosing **File | Save As,** and let them know how to name and where to save their files.

DEVELOP

Expect students at computers to spend about 30 minutes.

10. Assign students to computers and tell them where to locate **Circles All Around.gsp.** Distribute the worksheet. Tell students to work through step 11 and do the Explore More if they have time. Encourage students to ask their neighbors for help if they have questions about using Sketchpad.

11. Let pairs work at their own pace. As you circulate, here are some things to notice.

- In worksheet steps 2 and 3, encourage students to draw lots of circles and drag them. *How can you make the circle larger? How can you make it smaller? Can you drag the circle and keep it the same size?*

- In worksheet step 2, students may discover how to make concentric circles by first clicking the same center point each time they use the **Compass** tool. Introduce and define the term *concentric circles*, and then encourage students to draw some other circles as well.

- In worksheet steps 4–8, students use a different method to construct a circle. By selecting a segment and a point and choosing **Construct | Circle by Center+Radius**, students will construct a circle whose radius is determined by the segment. Students will likely find this construction very interesting because even though the segment isn't attached to the point, it changes the size of the circle when either endpoint is dragged.

- When a student pair has completed step 9, ask, *How does this new way of building a circle compare to using the Compass tool?* Here are some sample responses.

  They both make circles. And the circles all have center points.

  *This circle [**Compass** tool construction] comes with a point on it. When we drag the point, the circle changes size. This circle [**Circle by Center+Radius** construction] only has a point at the center. We can only change the size of the circle by changing the length of the segment.*

It seems like the length of this segment is equal to the radius of this circle [**Circle by Center+Radius** construction]. *We can drag the circle right onto the segment to show they are the same length.*

- In worksheet steps 10 and 11, students play games that require them to use and think about the **Circle by Center+Radius** command. For examples of what students will construct in each game, open **Circles All Around Present.gsp** and go to pages "Game 1" and "Game 2."

12. Encourage pairs that have played both games to move on and do the Explore More, worksheet steps 12–17.

13. If you have a flash drive available, collect sketches to display students' circle pictures from the shared computer. If students will print their work, model choosing **File | Print Preview**; in the dialog box that appears, set the image to fit on one page before clicking **Print.**

If students will save their work, remind them how now.

SUMMARIZE

Project the sketch. Expect to spend about 15 minutes.

14. Now that students are familiar with the radius of a circle, introduce the diameter. Open **Circles All Around Present.gsp.** Go to page "Summarize." Invite the class to observe.

- Select point A and choose **Transform | Mark Center.**

- Select $\overline{AB}$ and point B. Choose **Transform | Rotate.** In the window that pops up, enter 180 for the angle. Click **Rotate.**

- Observe that $\overline{AB}$ rotates halfway around the circle. Using the **Text** tool, label the rotated point as B'.

Sketchpad rounds measurements, so the measurement for the diameter may not always appear to be exactly twice the measurement of the radius.

- ***What do you think the length of this new segment BB' will be?*** Select points B and B' and choose **Measure | Distance** to check.

- Drag point B to change the size of the circle. As the circle changes size, ask students to check whether the length of $\overline{BB'}$ is always twice the length of $\overline{AB}$.

- ***The line segment through the center is called a diameter. What can you say about the diameter of a circle?*** [The diameter's length is twice the length of the radius.]

15. Have the class make a poster that shows an illustration of a circle with a radius, a diameter, and the circumference, and definitions for each term. Here are definitions to share.

   • *Circle*: the set of all points in a plane that are the same distance from a center point.

   • *Circumference*: the distance around a circle (the perimeter)

   • *Radius*: a line segment with its endpoints at the center of a circle and on the circle; the distance from the center of a circle to any point on the circle; half the diameter.

   • *Diameter*: a line segment that passes through the center of a circle and has its endpoints on the circle; the length of such a segment; twice the radius.

16. If time permits, discuss the Explore More. Have students share their designs and pictures. **Which circle construction method did you like best?** Let students share their favorite and the reason why.

EXTEND

1. Model drawing a circle using a string tied to a pencil and a tack.

   How does this method of constructing a circle compare to the way Sketchpad constructs circles? Take student responses. Here are sample responses.

   The string is like the segment when we use **Construct | Circle by Center+Radius.** *The string and the segment are the length of the radius.*

   With the **Compass** *tool, we drag the point on the circumference to change the size of the circle. It's like changing the length of the string.*

   It's easier to make a circle that's another size in Sketchpad. You just drag the point on the circumference made by the **Compass** *tool or the segment that controls the radius.*

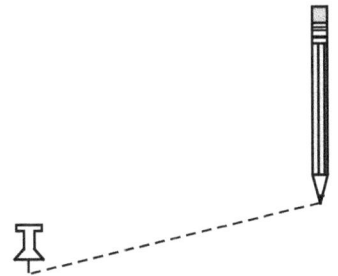

2. Students may wonder about line segments that "go across the circle but aren't radii or diameters." Open **Circles All Around Present.gsp** and go to page "Chord." Press *Animate Point B*.

Let students observe as point *B* travels once around the circle. Explain that the name for segment *AB* is *chord—a line segment with both endpoints on the circle.* Elicit ideas. ***When is AB longest?*** [When it is the diameter of the circle] With the animation stopped, you can also drag point *B* to facilitate discussion.

Circles All Around

For GSP5

Name:

Construct and explore circles.

EXPLORE

1. Open **Circles All Around.gsp.** Go to page "Compass Tool."

2. Use the **Compass** tool to construct several circles.

3. One by one, drag the circles and the points. Watch what happens.

Now you'll make a circle another way.

4. Go to page "Center+Radius".

5. Construct a segment.

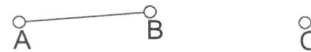

6. Construct a point not on the segment.

7. Label the points.

8. Select segment *AB* and point *C*
   (but not points *A* and *B*) and choose
   Construct | Circle by Center+Radius.

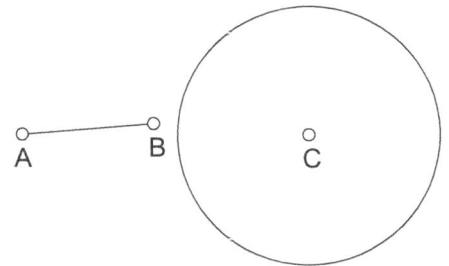

9. One by one, drag the segment, the circle,
   and each point. Watch what happens.

PLAY CIRCLE GAMES

10. Go to page "Game 1." Read the directions and play.

11. Go to page "Game 2." Read the directions and play.

EXPLORE MORE

12. Go to page "Explore More." Construct two points.

Circles All Around

continued

13. Select them and choose **Construct | Circle by Center+Point**. See what happens when you drag the points.

14. Draw four or five points on the screen.

15. Use the **Circle by Center+Point** command to construct some circles using these points.

Now you will color in some of your circles.

16. Select the circle and choose **Construct | Circle Interior**.

17. With the interior selected, choose **Display | Color** and pick a color.

New York City Title I Elementary School Activities with The Geometer's Sketchpad
© 2012 Key Curriculum Press

Grade 3 Activities

Place-Value Counter: Get to the Target

INTRODUCE

Project the sketch for viewing by the class. Expect to spend about 10 minutes.

1. Open **Place Value Counter Target.gsp.** Go to page "Counter." Tell students, *Today you're going to be solving puzzles that use this special counter.* Press the + button in the ones place repeatedly, pausing between presses, while the class observes. Press the + button in the tens place a few times; then the + button in the hundreds place. (Leave the rest of the buttons, including the − buttons, for students to investigate on their own.) Invite students' comments about the counter. Does it remind them of an odometer? Are there other places where they have seen a display that counts up (or down) using large numbers?

2. Press *Reset to 0* to clear the counter. *I want to reach 99. How can I do that?* Some students may suggest pressing buttons to "count by 10's to 90, and then count by 1's." *Is there another way we can reach 99 that requires fewer button presses?* [A more efficient approach, starting at 0, is to press the + button in the hundreds place once, and then press the − button in the ones place once.]

3. Distribute the worksheet. *Your goal is to reach each target number in as few button presses as possible.* Solve the first problem together. For each button pressed, record the updated number on the counter. Make sure students understand that this way of recording will let them come back and review the presses they used. In the case of 42, the recording should be written as 10, 20, 30, 40, 41, 42. There are lines for recording two trials for reaching each target. If students think that they can get to the target using fewer presses than they used the first time, they should try the problem again and list the numbers in the new solution. *At the end of each line, write the number of presses used.*

DEVELOP

Expect students at computers to spend about 30 minutes.

4. Assign students to computers and tell them where to locate **Place Value Counter Target.gsp.** Tell students to work on all the targets in steps 1–12 and do the Explore More if they have time.

5. Let pairs work at their own pace. As you circulate, observe and listen to students' conversations. Here are some things to notice.

   • In worksheet step 2, students may reach 48 using the sequence 10, 20, 30, 40, 41, 42, 43, 44, 45, 46, 47, 48. If so, challenge them to find a shorter sequence. Students may recall that reaching 99 in only two presses required them to jump past the target. They can use the same strategy here to obtain the shorter sequence 10, 20, 30, 40, 50, 49, 48.

- In worksheet step 2, when students have reached the target, ask, **Why did you jump to 40 for target 1, but jump to 50 for target 2?** Students may explain that 42 is closer to 40 than to 50, whereas 48 is closer to 50 than to 40.

- In worksheet step 3, students will likely realize that jumping to 60 (as opposed to 70) is the most efficient way to reach 61. The question is, what's the best way to jump to 60? Some students may only consider the sequence 10, 20, 30, 40, 50, 60. If so, ask, **Sometimes we've seen that it helps to jump past a number. Would it help to jump past 60?** Going to 100 and then back by 10's (100, 90, 80, 70, 60) produces a shorter sequence.

- When most of the class has completed worksheet step 4, consider checking in with the class. **You've reached targets that are between 1 and 99. What advice would you give someone for reaching a target in that range?** Take responses. One good idea is this student's: *First, check whether the number is closer to 0 or 100. If it's closer to 100, jump to 100. Otherwise, stay at 0. Then check which multiple of 10 is closest to the number. Jump to that number, moving either forward or backward by 10's. Then go either forward or backward by 1's to reach the target.*

- In worksheet steps 5 and 6, reaching the targets reveals an interesting property of target numbers ending in a 5. The standard rounding rule says that a number ending in a 5 should be rounded up. But if students round 15 up to 20, their sequence—10, 20, 19, 18, 17, 16, 15—takes seven presses. Rounding 15 down to 10 yields the sequence 10, 11, 12, 13, 14, 15—six presses. By comparison, in worksheet step 6, the fewest number of presses to get to 85 requires rounding 85 up to 90 first, yielding the sequence 100, 90, 89, 88, 87, 86, 85. Students can try reaching other two-digit targets ending in 5 to develop the idea that numbers closer to 0 (15, 25, 35, 45) should be rounded down to the nearest 10, whereas numbers closer to 100 (65, 75, 85, 95) should be rounded up to the nearest 10. In the case of 55, the number can be rounded either up or down; the number of presses is the same.

The rule works for all numbers between 1 and 99, except 51 through 55. There is no need, however, to raise these exceptions now unless a student does.

Students' strategies will likely be more tentative and incomplete. Give students plenty of time to experiment before expecting them to develop a general approach.

• In worksheet steps 7–10, students develop strategies for reaching three-digit numbers. One general approach is to round the target to the nearest thousand, then to the nearest hundred, and then to the nearest ten. For example, the target in step 10 is 782, which is 1000 when rounded to the nearest thousand, 800 when rounded to the nearest hundred, and 780 when rounded to the nearest ten. These three rounded numbers are landmarks to reach, as in this sequence: 1000, 900, 800, 780, 781, 782.

SUMMARIZE

Project the sketch. Expect to spend about 20 minutes.

6. Gather the class. Students should have their worksheets with them. Choose a problem from the worksheet and ask students to share the number of presses they used to get to the target. Facilitate discussion and have students come to the computer to share their solutions. Make sure students communicate the reasoning behind their presses, and highlight any common strategies presented if students don't point them out to their classmates.

7. To focus the discussion, you might ask questions such as these.

   How can we get to 99? How can we get to 999? How can we get to 9999?

   If the target is a one-digit number, how do you know whether your first jump should be to 10?

   If the target is a two-digit number, how do you know whether your first jump should be to 100?

   If the target is a three-digit number, how do you know whether your first jump should be to 1000?

   Can you give me several two-digit numbers, other than those on the worksheet, that you would reach most quickly by using only the + buttons? How can you tell by looking at the numbers that you'll use only the + buttons?

8. Solving the counter problems required no prior knowledge of rounding; the activity, in fact, helps to develop an intuitive sense of the logic behind the rounding rules. One rounding rule however does not fit so neatly into this activity. When a number ends in a 5, the standard rounding rule says to round up. For the targets in worksheet steps 5 and 6, however, students saw that when a number ended in 5, it did not always make sense to "round up" and jump to a higher number first.

EXTEND

What other questions about getting to targets using the counter occur to you? Here are sample student queries.

What two-digit number takes the most presses to reach?

What three-digit number takes the most presses to reach?

Suppose we weren't allowed to press certain buttons. How would that change things?

Suppose you weren't allowed to press a button two times in a row. How would that change things?

Can we predict correctly how many presses it will take?

Is there a way to predict whether a number will take the same number of presses whether you jump higher or lower?

ANSWERS

1. 10, 20, 30, 40, 41, 42

2. 10, 20, 30, 40, 50, 49, 48

3. 100, 90, 80, 70, 60, 61

4. 100, 90, 89, 88, 87

5. 10, 11, 12, 13, 14, 15

6. 100, 90, 89, 88, 87, 86, 85

7. 100, 110, 120, 130, 131, 132, 133, 134

8. 100, 200, 190, 180, 170, 171, 172

9. 1000, 900, 800, 810, 820, 830, 840, 839, 838, 837, 836

10. 1000, 900, 800, 790, 780, 781, 782

11. 1000, 2000, 3000, 4000, 3900, 3800, 3700, 3600, 3610, 3620, 3619

12. 10000, 9000, 8000, 7990, 7980, 7970, 7969, 7968

13. Answers will vary.

Get to the Target

For
GSP5 Name:

Use what you know about place value to reach the targets.

EXPLORE

Open **Place Value Counter Target.gsp.** Go to page "Counter."

Get to each target in the fewest number of presses possible.

Record the number that each press takes you to and the number of presses used.

Always press *Reset to 0* to start, so you begin at 0.

1. The target is 42.

_____

_____

2. The target is 48.

_____

_____

3. The target is 61.

_____

_____

4. The target is 87.

_____

_____

5. The target is 15.

_____

_____

New York City Title I Elementary School Activities with The Geometer's Sketchpad
© 2012 Key Curriculum Press

75

6. The target is 85.

_____

_____

7. The target is 134.

_____

_____

8. The target is 172.

_____

_____

9. The target is 836.

_____

_____

_____

10. The target is 782.

_____

_____

_____

11. The target is 3,619.

_____

_____

_____

_____

12. The target is 7,968.

_____

_____

_____

_____

EXPLORE MORE

13. Make up some targets. Reach them in the least number of presses possible. Then challenge a partner to reach them.

_____

_____

_____

_____

Factor Puzzles: Number Sense and Logical Reasoning

INTRODUCE

Project the sketch on a large-screen display for viewing by the class. Expect to spend about 10 minutes.

1. Open **Factor Puzzles.gsp.** Go to page "Puzzles."

2. Familiarize students with the factor-puzzle rules described on the screen. Two points are worth highlighting.

   • A product will appear only when exactly two letters are on the right of the divider line. Thus, if students are viewing the product of b and c and then drag d to the right of the line, the product of b, c, and d will not appear.

   • The same letter can be used in more than one product. For example, if students are viewing the product of b and c, they can drag b back to the left of the divider and then drag d across in its place to view the product of d and c.

3. Solve one puzzle together as a class. Invite students to pick letter pairs to drag across the line. Have them record the products. Students should offer solutions to the puzzle, but don't spend too much time on a discussion. Students will have a chance to develop their own solution strategies when they work alone or in pairs at computers.

4. Briefly discuss ways to keep track of the products on paper. One possibility is to write statements like $a \times b = 15$ or $b \times c = 10$. Another option is to make a table like this one.

Letters				Product
a	b	c	d	
✓	✓			15
	✓	✓		10

5. Explain that students can go on to page "More Puzzles" when they are ready. On that page, the possible values for each letter range from 1 through 14.

DEVELOP

Expect students at computers to spend about 30 minutes.

6. Let students know where they will locate **Factor Puzzles.gsp** when they work independently. Have pairs or individuals solve the number puzzles, working at their own pace.

7. If you have the opportunity to observe as students solve the puzzles, notice the strategies they employ. Make note of any students whose strategies you may want to share in a small group or class discussion. Some examples of students' mathematical thinking follow.

   • While some students may find it sufficient to list only the products of various letter pairs, others may also need to list the possible factors for each product.

   • Students may not realize that if $a \times b = 21$, for example, there are two possible values for a and b: Either $a = 3$ and $b = 7$, or $a = 7$ and $b = 3$. This is because multiplication is *commutative* ($3 \times 7 = 7 \times 3$). To determine which pair of values solve the puzzle, students will need to check other letter pairs to gather more information.

   • Students will likely discover that it is helpful to use the same letter in more than one factor pair. Suppose that $a \times b = 30$. Because a and b are no larger than 9, either $a = 6$ and $b = 5$, or $a = 5$ and $b = 6$. To choose between these two possibilities, use b in another factor pair. If $b \times c$ equals 24, for example, b must equal 6 because 5 is not a factor of 24.

   • On page "More Puzzles," the possible values for each letter range from 1 through 14. With more numbers to choose from, students will sometimes need to dig deeper to find the values of the factors. If students discover that $b \times d = 56$ and $a \times d = 28$, for example, then d might be either 7 or 14. Checking other letter pairs will enable students to determine which possibility is correct.

SUMMARIZE

Expect to spend about 15 minutes on this part of the activity. Plan additional time for step 10.

8. After some students have solved factor puzzles, provide an opportunity for them to describe their strategies to each other. Explain that the group should focus on the reasoning behind the strategies.

9. Ask students to share any of the puzzles that they felt especially proud to have solved and to explain what made them challenging.

10. ***What other questions can you ask that you may or may not be able to answer?*** Encourage all student inquiry. Mathematical questions of interest include these.

- What secret values require that you know the fewest products? The most products?

- What if Sketchpad showed the product only if you moved three variables across the line?

- What if you had five secret numbers? How many pairs would there be?

EXTEND

1. Pairs of students can create factor puzzles for each other using page "Make Your Own." The directions for making a puzzle are on that page.

2. Explain, ***I was working on a factor puzzle and wrote down the product of four pairs of letters: ab, bc, cd, and ad. I found it really interesting that even with all this information, I wasn't able to solve the puzzle! Can you come up with a puzzle like that?*** [If, for example, the product of each letter pair is 36, then *a* through *d* could all equal 6, or *a* and *c* could equal 4 with *b* and *d* equalling 9.]

3. Discuss how many different pairs of letters students could choose from when solving the puzzles. [There are six in total: *ab, ac, ad, bc, bd,* and *cd.*] Because multiplication is commutative, $a \times b$ is the same as $b \times a$: Both represent the product of *a* and *b*.]

4. Optionally, discuss how the letters *a, b, c,* and *d* serve as *variables* in the factor puzzles. Each letter represents a number, but this numerical value changes from puzzle to puzzle.

INTRODUCE

Project the sketch for viewing by the class. Expect to spend about 10 minutes.

1. Open **Jump Along Multiplication.gsp.** Go to page "Jump Along." Ask students to describe what they see. A rabbit is sitting at 0 on a number line.

2. Tell students, **Watch what happens when I press Jump Along.** Press *Jump Along.* **What did the rabbit do?** Take responses. If necessary, press *Move Rabbit to 0* and then *Jump Along* so that students can watch the movement again. The rabbit makes four jumps of size 1 and lands at 4.

Although it is possible to enter a non-integer value for the *Number of Jumps* or *Jump By* parameter, students will use only integer values in this activity.

3. Change the *Jump By* number to 2 by double-clicking *Jump By* with the **Arrow** tool, entering 2, and pressing **OK.** Reset the rabbit by pressing *Erase Traces* and *Move Rabbit to 0.* Press *Jump Along.* Ask students to describe what the rabbit did. Make sure everyone sees that *Number of Jumps* tells them how many times the rabbit jumped and *Jump By* tells the size of each jump.

4. Distribute the worksheet. Together, look at the table in step 2. Ask students to fill in all but the last column of row a. Students should relate this row to the rabbit in the sketch: The rabbit made four jumps of size 2 and landed at 8. Point out that the rabbit always starts at 0.

   Let's look at the last column of the table. The rabbit landed at 8. Can you write a number sentence that explains how the rabbit made its way to 8? As a class, develop the number sentence $2 + 2 + 2 + 2 = 8$. **Why are we adding 2 each time?** [The size of the rabbit's jump was always 2.] **Why do we add four 2's?** [The rabbit jumped 4 times.]

As students solve more problems, they will make the connection between repeated addition and multiplication.

5. Write 4 on the board and ask, **Suppose I want to write a different number sentence that starts with 4 and means the same thing as** $2 + 2 + 2 + 2 = 8$. **How can we complete this number sentence without using addition signs?** Students may suggest $4 \times 2 = 8$. **How does this multiplication sentence relate to the rabbit's movement along the number line?** [The rabbit made 4 jumps, each of size 2, and landed at 8.]

This is correct because of the *commutative property of multiplication.*

Note that students may also suggest $2 \times 4 = 8$ as a valid multiplication statement. Explain, **That is certainly another correct way to represent** $2 + 2 + 2 + 2 = 8$. **Let's agree as a class to write the number of jumps first.** Establishing this convention with students will facilitate conversations at computers and class discussions.

DEVELOP

Expect students at computers to spend about 30 minutes.

6. Assign students to computers and tell them where to locate **Jump Along Multiplication.gsp.** Tell students to work through steps 1–4 and do the Explore More tasks if they have time. Encourage them to ask a neighbor for help if they have questions about using Sketchpad.

7. Let pairs work at their own pace. As you circulate, listen to students' conversations. Here are some things to notice.

   • In worksheet step 2, students do not need to predict where the rabbit will land before entering the numbers and pressing *Jump Along.* The goal is for students to make a connection between the visual representation of the rabbit's jumps and the number sentences that describe how the rabbit jumped and where it landed.

   • To start a new problem in worksheet step 2, students should press *Erase Traces* and *Move Rabbit to 0.* For variety, students can change the color of the rabbit's traces by selecting the point below the rabbit and choosing **Display | Color** to pick a new color.

   • If students happen to enter numbers that cause the rabbit to jump beyond the visible portion of the screen, the extra jumps will not be visible if students scroll across after the rabbit has finished jumping.

   • In worksheet step 3, if any students are asking for help, prompt them to persist in doing their own thinking by asking, *Have you experimented? Are you thinking about all the information given? What piece of information about the rabbit are you missing? Do you know how many jumps the rabbit takes? Do you know the size of each jump? Do you know where the rabbit lands?*

SUMMARIZE

Project the sketch. Expect to spend about 20 minutes.

8. Gather the class. Students should have their worksheets with them. Go to page "Jump Along." Discuss worksheet steps 3a and 3b. Use the model to display both five jumps of 2 and two jumps of 5 on the number line. Make the traces different colors.

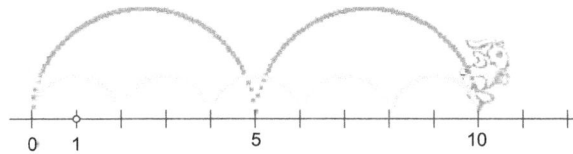

Number of Jumps = 2   Jump Along

Jump By = 5

What is the same and what is different about these trips the rabbit has taken? Take responses. Sample responses follow.

The numbers we put in are the same—2 and 5—but they're reversed.

The actions that the rabbit takes are related. In one problem, the rabbit takes five jumps of 2. In the other problem, the rabbit takes two jumps of 5.

The rabbit landed at the same place in both problems.

The number sentences look almost alike: $2 \times 5 = 10$ *and* $5 \times 2 = 10$.

It doesn't matter what order you multiply 2 and 5 in. The answer is the same.

Whenever you multiply two numbers, it doesn't matter what order you multiply them in. The product is always the same.

9. Write on the board the equation $6 \times 3 = ?$. ***Think of a story problem for the rabbit that goes with this number sentence.*** One possibility is this: The rabbit starts at 0 and takes six jumps of 3. Where does it get to?

   As another example, write the equation $4 \times ? = 20$. A story problem is this: The rabbit starts at 0. It wants to get to 20 in four jumps. It wonders how big each jump will need to be.

EXTEND

1. For students who are ready to work with larger numbers, you can rescale the number line by dragging the point sitting at 1 so that it is closer to 0.

2. Use the number line model to introduce students to the distributive property.

   • Set the rabbit to start at 0 and make 4 jumps of 3.

   • Record "4 jumps of 3" as well as 4×3 on the board.

   • With the rabbit at 12, take 2 more jumps of 3.

- Record "2 jumps of 3" as well as 2 × 3 on the board.

- Have students look at the trace of the rabbit's entire trip, and observe that altogether the rabbit took six jumps of 3, landing at 18. Elicit the idea that six jumps of 3 is another way of viewing four jumps of 3 plus two jumps of 3.

- Express this symbolically as 6 × 3 = 4 × 3 + 2 × 3.

3. **What other questions can you ask about jumping along the number line?** Encourage student curiosity. Here are some sample student queries.

   How many different ways can the rabbit jump to 11, starting at 0?

   What if our number sentence is 3 × 2 + 1 × 4 = 10? *What story about the rabbit could go along with this number sentence?*

   What if the Jump By *number is less than 0? What happens then?*

ANSWERS

2.

The Rabbit Makes...	Number of Jumps	Jump By	Get To	Multiplication Number Sentence
a. 4 jumps of 2	4	2	8	4 × 2 = 8
b. 3 jumps of 4	3	4	12	3 × 4 = 12
c. 5 jumps of 2	5	2	10	5 × 2 = 10
d. 6 jumps of 3	6	3	18	6 × 3 = 18

3.

Problem	Number of Jumps	Jump By	Get To	Multiplication Number Sentence
a.	5	2	10	5 × 2 = 10
b.	2	5	10	2 × 5 = 10
c.	4	3	12	4 × 3 = 12
d.	3	4	12	3 × 4 = 12
e.	4	6	24	4 × 6 = 24
f.	6	4	24	6 × 4 = 24

4. Rows a and b have the same numbers, only "flipped." Row a shows that five jumps of size 2 land the rabbit at 10. Row b shows that two jumps of size 5 also land the rabbit at 10. The same relationship can be found between rows c and d and between rows e and f.

5. The rabbit can take five jumps to land on 20.

6. The rabbit can take four jumps to land on 28.

7. The rabbit can jump by 6 to land on 36.

8. The rabbit cannot reach the odd numbers.

9.

The Rabbit Makes . . .	Number of Jumps	Jump By	Get To	Number Sentence
a. 3 jumps of 2	3	2	8	$2 + 3 \times 2 = 8$
b. 5 jumps of 4	5	4	22	$2 + 5 \times 4 = 22$
c. 7 jumps of 3	7	3	23	$2 + 7 \times 3 = 23$

Jump Along

For GSP5

Name:

Help a rabbit jump along a number line.

EXPLORE

1. Open **Jump Along Multiplication.gsp.** Go to page "Jump Along."

2. Get the rabbit where it wants to go. Use the table to record.

Rabbit Starts at 0

The Rabbit Makes. . .	Number of Jumps	Jump By	Get To	Multiplication Number Sentence
a. 4 jumps of 2				
b. 3 jumps of 4				
c. 5 jumps of 2				
d. 6 jumps of 3				

3. Get the rabbit where it wants to go. Use the table on the next page to record.

   a. The rabbit wants to jump by 2's and get to 10.

   b. The rabbit wants to take 2 jumps and get to 10.

   c. The rabbit wants to jump by 3's and get to 12.

   d. The rabbit wants to take 3 jumps to get to 12.

   e. The rabbit wants to jump by 6's to get to 24.

   f. The rabbit wants to take 6 jumps to get to 24.

New York City Title I Elementary School Activities with The Geometer's Sketchpad
© 2012 Key Curriculum Press

Jump Along

continued

Rabbit Starts at 0

The Rabbit Makes. . .	Number of Jumps	Jump By	Get To	Multiplication Number Sentence
a.				
b.				
c.				
d.				
e.				
f.				

4. Look at your table in step 3. What patterns can you find?

_____

_____

_____

_____

EXPLORE MORE

5. How close can the rabbit get to 21 when it jumps by 4?

_____

6. How close can the rabbit get to 30 when it jumps by 7?

_____

7. How close can the rabbit get to 35 when it makes 6 jumps?

_____

8. If the rabbit jumps by 2's, what numbers is it not able to get to?

_____

Jump Along

continued

9. Go to page "Explore More."

   Drag the point below the rabbit so that it moves to 2.

   Show each problem on the number line. Record what happens.

Rabbit Starts at 2

The Rabbit Makes. . .	Number of Jumps	Jump By	Get To	Number Sentence
a. 3 jumps of 2				
b. 5 jumps of 4				
c. 7 jumps of 3				

Jump Along:
Factor Families on the Number Line

For GSP5

ACTIVITY NOTES

INTRODUCE

Project the sketch for viewing by the class. Expect to spend about 10 minutes.

1. Open **Jump Along Factor Families.gsp.** Go to page "Jump to 12." Ask students to describe what they see. A rabbit is sitting at 0 on a number line.

2. Tell students, ***Watch what happens when I press*** **Jump Along.** Press *Jump Along.* ***What did the rabbit do?*** Take responses. If necessary, press *Jump Along* again so that students can watch the movement again. The rabbit makes 4 jumps of size 1 and lands at 4.

3. Press *Erase Traces* to clear the path of the rabbit's jumps.

Although it is possible to enter a non-integer value for the *Number of Jumps* or *Jump By* parameter, students will use only integer values in this activity.

4. Change the *Jump By* number to 2 by double-clicking *Jump By* with the **Arrow** tool, entering 2, and pressing OK. (Alternatively, you can select *Jump By* with the **Arrow** tool and press either + or − on the keyboard to increase or decrease the value.)

Press *Jump Along.* Ask students to describe what the rabbit did. Make sure everyone sees that *Number of Jumps* tells them how many times the rabbit jumped and *Jump By* tells the size of each jump.

5. Distribute the worksheet. Point to 12 on the number line. ***How can the rabbit get to 12 if it starts at 0?*** Take a suggestion and model it with the rabbit. Have students describe the rabbit's movement using language such as, "The rabbit made 6 jumps of size 2 and landed at 12."

6. ***Let's change the color of the jumps each time.*** Select the point below the rabbit and choose **Display | Color** to pick a different color. ***Now, when we press*** **Jump Along,** ***the rabbit's jumps will be a new color.*** Ask students for a different pair of numbers that will land the rabbit at 12, and try the numbers using the new color. ***When you work with the sketch, change the color each time you try a different way to jump to the target.***

7. Together, look at the table in worksheet step 2. Model how to fill in the first row using one of the jumps to 12 you made as a class. ***Let's look at the last column of the table. Can you write a multiplication sentence that explains how the rabbit made its way to 12?*** As a class, develop the multiplication sentence (for example, $6 \times 2 = 12$). ***Why are we multiplying 6 by 2?*** [There were 6 jumps of size 2.]

This is an example of the *commutative property* of multiplication.

Note that students may also suggest $2 \times 6 = 12$ as a valid multiplication statement. Explain, ***That is certainly another correct way to represent 6 jumps of size 2. Let's agree as a class to write the number of jumps first.*** Establishing this convention with students will facilitate conversations at computers and in class discussions.

8. If you will have students print their sketches, model how to choose **File | Print Preview,** and in the dialog box that appears, set the image to fit on one page and then click Print.

DEVELOP

Expect students at computers to spend about 30 minutes.

9. Assign students to computers and tell them where to locate **Jump Along Factor Families.gsp.** Tell students to work through steps 1–7. Encourage students to ask a neighbor for help if they have questions about using Sketchpad.

   Explain that in worksheet step 2, everyone will try to reach the target number of 12. In worksheet steps 4 and 6, however, students will choose a target from the list of numbers provided.

10. Let pairs work at their own pace. As you circulate, listen to students' conversations. Here are some things to notice.

   • In worksheet steps 2, 4, and 6, students should change the color of the rabbit's traces each time. Tell students to check the *Jump By* and *Number of Jumps* numbers carefully before pressing *Jump Along.* If students make an error and pick numbers that are not factors of the target number, they will not be able to erase the trace of just the last jump. *Erase Traces* deletes all traces.

   • If students happen to enter numbers that cause the rabbit to jump beyond the visible portion of the screen, the extra jumps will not be visible if students scroll across the screen after the rabbit has finished jumping.

   • Traces will not be saved when students go to another page of the sketch or if they save their sketch. If you are having students print their sketches, be sure they print the current page before moving on to the next target number.

   • Ask, *How do you know when you've found all the ways to reach the target number?* Some students will be systematic; others will have a more informal approach.

   We started with jumps of size 1; they always work. Then we thought about jumps of size 2, then jumps of size 3, and so on.

   We just tried different numbers until we couldn't find anymore.

New York City Title I Elementary School Activities with The Geometer's Sketchpad
© 2012 Key Curriculum Press

We tried numbers starting from 1. We knew 1 × 12 worked, so we also knew 12 × 1 worked. We kept going until the number sentences started to repeat. For example, we found 3 × 4 = 12, so we knew 4 × 3 = 12 worked. Our next number to try was 4, but we already had it in 4 × 3 = 12. We knew there weren't any other ways. Every number in our Number of Jumps column of the table also appeared in the Jump By column.

SUMMARIZE

Project the sketch. Expect to spend about 20 minutes.

11. Gather the class. Students should have their worksheets with them. Go to page "Jump to 12." Begin by presenting an example in order to elicit students' understanding of the *commutative property of multiplication.* (You may or may not wish to mention this term.) Use the model to display 2 jumps of size 6. **How does knowing 2 jumps of size 6, or 2 × 6 = 12, help you find another way to reach 12?** Take responses. Have a volunteer come up and model 6 × 2 = 12, using a different trace color. Sample responses follow.

    You can put in the same numbers but reverse them—6 jumps of size 2.

    The actions that the rabbit takes are related. Two jumps of size 6 is related to 6 jumps of size 2.

    It doesn't matter what order you multiply 2 and 6 in. The answer is the same; so 2 jumps of size 6 and 6 jumps of size 2 will both land at 12.

    Whenever you multiply two numbers, it doesn't matter what order you multiply them in. The product is always the same.

12. Have volunteers come up to the computer and finish the remaining ways for the rabbit to land at 12, using a different color for each way. **What patterns did you notice?** Students may make the following observations.

    When the size of the jump increased, it took fewer jumps to get to 12.

    The opposite was true, too. When the size of the jump got smaller, it took more jumps to get to 12.

    All of the jumps the rabbit took fit under the biggest jump of all—1 jump of size 12.

    The pattern made by the jumps is symmetric. If I drew a vertical line through the number line at 6, I could reflect one half of the pattern across the line to show the other half.

The rabbit traced 1 half circle in orange, 2 half circles in pink, 3 half circles in light purple, 4 half circles in dark purple, 6 half circles in light blue, and 12 half circles in red. These numbers are all factors of 12.

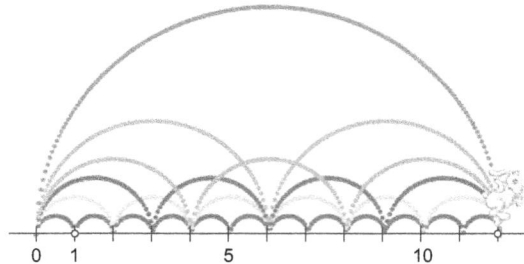

13. **How can you list the factors of 12 just by looking at your table?**
Students can list all of the factors of 12 (1, 2, 3, 4, 6, and 12) just by reading the numbers in either of the first two columns of their table.

14. Ask students to share their results from worksheet steps 4 and 6. In particular, ask them how the target numbers in step 6 differ from the target numbers in step 4. Sample responses follow.

There are only two ways to get to each of the targets in step 6.

In step 6, the rabbit can either jump by 1, or it can jump the whole way in one jump. So, for 13, it's either 13×1 or 1×13.

The targets in step 4 have lots of factors. The numbers in step 6 only have two factors. One of those factors is 1, and the other is the number itself.

The targets in step 6 are prime numbers.

EXTEND

15. For students who are ready to work with larger numbers, rescale the number line by dragging the point at 1 so that it is closer to 0. With more of the number line visible, students can reach targets larger than 25.

16. **What other questions can you ask about jumping along the number line?** Encourage student curiosity. Here are some sample student queries.

How many different ways can the rabbit jump to 100, starting at 0?

Do you think there are more ways to reach large target numbers than smaller target numbers?

What target number under 100 has the most number of ways for the rabbit to jump?

What if the Jump By number is less than 0? What happens then?

Is it possible that there's only one way for a rabbit to jump to a particular target number?

I noticed that for most targets, the rabbit could reach them in an even number of ways. There are eight ways to reach 24, six ways to reach 20, four ways to reach 10, and so on. Why is that? Are there any target numbers that can be reached in an odd number of ways?

ANSWERS

2.

Number of Jumps	Jump By	Multiplication Number Sentence
1	12	$1 \times 12 = 12$
2	6	$2 \times 6 = 12$
3	4	$3 \times 4 = 12$
4	3	$4 \times 3 = 12$
6	2	$6 \times 2 = 12$
12	1	$12 \times 1 = 12$

3. Answers will vary. Students may make the following observations.

   • The larger the *Jump By* number, the smaller the number of jumps, and vice versa: the smaller the *Jump By* number, the larger the number of jumps.

   • Students may notice the commutative property of multiplication at work. For example, if 3 jumps of 4 works, then so will 4 jumps of 3.

4. Answers will vary. Check students' tables.

5. Answers will vary. See observations listed in answer for 4.

6. Answers will vary. Check students' tables.

7. Answers will vary. Students should notice that there are only two ways for the rabbit to reach the target number.

Jump Along Factor Families   For GSP5   Name:

Find all the ways a rabbit can jump to different targets.

EXPLORE

1. Open **Jump Along Factor Families.gsp.** Go to page "Jump to 12."

2. The rabbit wants to get to 12. How many different ways can it get there?

   Double-click the *Number of Jumps* and *Jump By* numbers to change them. Then press *Jump Along* to start the rabbit.

   Make each set of jumps a different color. Select the point below the rabbit. Then pick a new color by choosing **Display | Color.**

   Use the table to record the different ways.

   Target Number: 12

Number of Jumps	Jump By	Multiplication Number Sentence

New York City Title I Elementary School Activities with The Geometer's Sketchpad
© 2012 Key Curriculum Press

3. Look at the jumps traced by the rabbit. What patterns do you see?

_____

_____

_____

4. Go to page "Jump to Target."

   Pick a target from this list: 16, 18, 20, 24.

   How many different ways can the rabbit get to the target?

   Make each set of jumps a different color.

   Use the table to record the different ways.

Target Number: _____

Number of Jumps	Jump By	Multiplication Number Sentence

5. Look at the jumps traced by the rabbit. What patterns do you see?

_____

_____

_____

6. Pick a target from this list: 13, 17, 19, 23.

How many different ways can the rabbit get to the target?

Make each set of jumps a different color.

Use the table to record the different ways.

Target Number: _____

Number of Jumps	Jump By	Multiplication Number Sentence

7. Look at the jumps traced by the rabbit. How are the jumps for this target number different?

_____

_____

_____

New York City Title I Elementary School Activities with The Geometer's Sketchpad
© 2012 Key Curriculum Press

Comparing Fractions:
Number Sense and Benchmarks

INTRODUCE

Project the sketch for viewing by the class. Expect to spend about 10 minutes.

1. Distribute the worksheet and ask students to work on comparing the fractions in step 1 for a few minutes. (For now, they should ignore the write-on line beside each pair of fractions.) Explain that this is a warm-up for the exploration the class will do together. It's not expected that students will be able to compare all pairs.

2. Open **Comparing Fractions Number Sense.gsp** and go to page "Compare." To begin, make sure the model looks like the illustration at right. If students have not used a fraction-circle model before, ask questions to check that students are able to explain the relationships between the numerator of a fraction, its denominator, the number of parts in the circle, and the parts shaded. You might ask, *What would the top circle look like if I changed 1/4 to 1/8?* Take responses and then change the denominator of the top fraction by selecting the denominator with the **Arrow** tool and pressing **+** on the keyboard four times. As you do, let students know how you are changing the parameter for the denominator, and pause between each press.

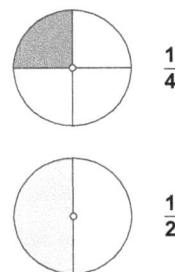

$\frac{1}{4}$

$\frac{1}{2}$

DEVELOP

Continue to project the sketch. Expect to spend about 35 minutes.

3. Facilitate discussion of the fraction pairs in steps 1a–g. Here are suggestions for using the model. Because visual models are being presented, make sure students know when the *numerical* representation is being discussed. Students are asked to write pairs of fractions on the worksheet. This provides an opportunity to check that students are connecting the visual examples with their fraction number sense and with the numerical representation of a fraction.

A. 2/9 AND 5/9

Using the model, show 2/9 in both circles and then increase the numerator to 5 in one fraction by selecting the numerator and pressing **+** three times. Pause between each press so that students can see the effect on the circle. *What can you say about comparing fractions that have the same denominator?*

Comparing Fractions: Number Sense and Benchmarks

continued

You may wish to introduce the term *common denominators* now and use it in this activity.

When the denominators are the same, it's easy to compare the fractions. The parts are the same size, so the fraction representing more parts—the fraction with the larger numerator—is greater. With the numerator 5 selected, press + until the numerator is 8 or 9 to further illustrate this.

In worksheet step 1a, have students write another pair of fractions that have the same denominator and different numerators. Then have students use the symbols for greater than and less than to show which fraction is greater.

B.   4/7 AND 4/11

Using the model, show 4/7 in both circles and then, in one fraction, increase the denominator to 11 by selecting the denominator and pressing + four times. Pause between each press so that students can see the effect on the circle. *What can you say about comparing fractions that have the same numerator?*

When the numerators are the same, it is also easy to compare the fractions. Each fraction has the same number of equal parts, but the size of the parts differs. The more parts, the smaller the parts are. The fraction with the larger denominator is less than the fraction with the smaller denominator.

Exploring extreme cases is worthwhile if the result is an engaging and memorable demonstration of an important concept.

Students may propose using the model to increase the denominator to even larger numbers. Select the larger denominator and, pressing the + key repeatedly, go all the way to 50.

On the worksheet, students should write another pair of fractions that have the same numerator and different denominators, and indicate which fraction is greater.

C.   2/4 AND 5/10

A denominator of 100 makes it impossible to see the divisions (which students enjoy seeing), whereas a denominator of 50 allows students to discern the divisions.

Students should recognize that both of these familiar fractions are equivalent to 1/2. Promote discussion about fractions that are equivalent to 1/2. Test a number of cases using the model, allowing time for students to conclude that for fractions equivalent to 1/2, the denominator is twice the numerator. Students may again suggest trying some very large values for the denominator. The quick way to change a parameter is to double-click the number, enter a new value in the dialog box that appears, and click OK. Use this method when you aren't interested in showing incremental changes.

Students should write another pair of fractions that are equivalent to 1/2, and, therefore, to each other, and use the equal sign to show the relationship.

D. 9/16 AND 4/9

Neither the numerators nor the denominators are the same. Another strategy is needed. A sample student response is this one: *I know that 4/9 is less than 1/2 because 4 is less than half of 9, and I know that 9/16 is larger than 1/2 because 9 is more than half of 16. So, 9/16 is greater.* If no student suggests a successful strategy, start the class off by asking whether 4/9 is greater than or less than 1/2.

When ideas have been considered, model 4/9 in one fraction circle by changing the numerator to 1 and the denominator to 9. Then select the numerator and press + until the numerator reaches 4. Stop and ask for students' thinking. ***Are you convinced that 4/9 is less than 1/2? What will the model look like if I press + again?*** Press + again to show 5/9. Then press − to return to 4/9.

Model 9/16 in the other fraction circle, changing the numerator to 1 and the denominator to 16 to start. Select the numerator and press + until the numerator is 9. Again discuss the results.

Students should write another pair of fractions for which relating the fractions to 1/2 is a handy way to compare them. Invite a few students to share the fraction pairs they have written, and model the fractions. (Some students may write pairs of fractions for which thinking about equivalence is an equally handy strategy, for example, 2/5 and 7/10. If students propose this alternate strategy, take time to discuss it.)

E. 2/21 AND 7/8

Thinking about how close the fractions are to benchmarks is again useful. This time students may compare the given fractions to the benchmarks 0 and 1, explaining that 2/21 is close to 0, and 7/8 is close to 1. Ask students to explain how they know that 2/21 is close to 0, and how they know 7/8 is close to 1. ***What can you say about fractions that are close to zero?*** [The numerator is much smaller than the denominator.] ***What can you say about fractions that are close to 1?*** [The numerator is close to the denominator.]

In this model, increasing the numerator until the fraction is greater than 1 will not result in the corresponding picture. Go to page "Improper Fractions" to work with fractions greater than 1.

Show 21/21 in one fraction circle, select the numerator and press −
repeatedly until the circle shows 2/21. In the other circle, set the fraction
to 1/8 and increase the numerator repeatedly until it shows 7/8.

Students should write another pair of fractions for which thinking
about the benchmarks 0 and 1 is useful. Invite students to share their
fraction pairs.

F.   3/4 AND 5/6

Some students may propose that the fractions are equivalent because
each is "*missing one part.*" Show the fractions in the model by double-
clicking each parameter and changing its value in the dialog box.
Students will see that the fractions are not equivalent. ***Does that make
sense? Why aren't the fractions equal to each other? After all, each
whole is missing exactly one part.***

Direct students' attention to the model and ask them to observe just one
of the fraction circles. Select *both* the numerator and the denominator
and press **+** on the keyboard, causing the numerator and denominator
to increase by one simultaneously. Repeat several times, pausing after
each press. ***What can you say?*** Note whether students are convinced that
as the number of parts increases, the size of the missing part decreases.

Elicit the idea that when the missing part is bigger, the given fraction is
smaller. Students should reason that 5/6 is greater than 3/4 because 5/6
is 1/6 less than 1, whereas 3/4 is 1/4 less than 1.

Students should write another pair of fractions for which thinking
about the "missing part" is useful.

G.   8/8 AND 9/13

Students can compare again using the benchmark 1. ***What can
you say about fractions equal to 1?*** [The numerator is equal to the
denominator.] Students should recognize that 8/8 is equal to 1 and 9/13
is less than 1.

Students may want to model, using the sketch, other pairs in which one
fraction is equal to 1 and the other is less than 1.

4. To include discussion of fractions greater than 1, go to page "Improper
   Fractions." Make sure the models show 1/6 and 1/8 to start. Write the
   fractions 9/6 and 16/8 on the board, and ask students to think about
   how they compare. Provide thinking time.

What do you think the model will look like if we show 9/6 and 16/8? Select the numerator in 1/6 and press + until 9/6 is shown. Select the numerator in 1/8 and press + until 16/8 is shown. *How can you tell that a fraction is greater than 1?* [The numerator is greater than the denominator.] *How can you tell that a fraction is between 1 and 2?* [The numerator is greater than the denominator, but less than twice the denominator.]

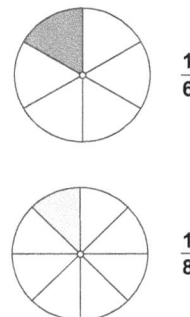

$\frac{1}{6}$

$\frac{1}{8}$

If time allows, have students suggest pairs of fractions to compare in which at least one of the fractions is greater than 1. Discuss students' thinking about the comparisons, and model the fractions.

SUMMARIZE

Working with or without the projected sketch, expect to spend about 15 minutes.

5. Point out that students have devised a number of strategies for comparing fractions. Focus the discussion on the usefulness of looking for a handy, or efficient, strategy when comparing fractions. Ask students to take a few minutes to consider the fractions in worksheet step 2. *Think about which strategy helps you compare the two fractions in each pair most easily and quickly.*

When most students have had a chance to consider all the pairs, ask about each pair, inviting students to share the strategies they found easiest to use.

(Note that, in general, students should not think that they need to agree on a most efficient strategy when thinking about numbers and computation; what is most efficient depends upon an individual's number sense and experience with computation.)

EXTEND

1. Have students work independently in pairs using pages "Compare" and "Improper Fractions." Give these directions:

   • One student chooses a pair of fractions to compare, trying to stump the other. The student must know which fraction is greater.

   • The second student chooses the fraction he or she thinks is greater.

   • The partners explain their reasoning and model the fractions to check.

   • Students can keep score on paper.

For GSP5   ACTIVITY NOTES

2. Page "Match Up" provides another model for independent practice comparing fractions. The sketch randomly generates pairs of fractions and shows them numerically and in fraction-circle models. Students should determine which fraction is greatest by using the numerical representation as well as by deciding which fraction-circle model shows which fraction. For a hint, students can press *Show Lines.* They can also compare the fraction circles directly by selecting the center of one circle and dragging it onto the center of the other circle.

3. Present three or more fractions and ask students to put them in order. Some students may find it helpful to use the model.

Compare the Fractions

For GSP5 Name:

Find handy ways to compare fractions.

EXPLORE

1. Compare the fractions in each pair. Use the symbols >, <, or = to show how they compare.

   a. 2/9   5/9   _____

   b. 4/7   4/11   _____

   c. 2/4   5/10   _____

   d. 9/16   4/9   _____

   e. 2/21   7/8   _____

   f. 3/4   5/6   _____

   g. 8/8   9/13   _____

2. For each pair of fractions, think about which strategy helps you compare the fractions quickly and easily.

   a. 6/11   24/50

   b. 8/9   5/6

   c. 3/12   8/12

   d. 30/60   7/14

   e. 40/50   4/4

   f. 7/17   7/100

   g. 11/12   2/15

Dividing and Subdividing:
Fractions on the Number Line

INTRODUCE

Project the sketch for viewing by the class. Expect to spend about 10 minutes.

1. Open **Dividing and Subdividing.gsp.** Go to page "a." Explain, *Today you're going to use some special Sketchpad fraction tools to find the locations of fractions between 0 and 1 on a number line.* Point out the four identical number lines. Explain that the class will be working with only the part of the lines from 0 to 1. Press *Hide Lines* so that only that part remains.

   Let's see how we can use our fraction tools to find $\frac{1}{4}$. We'll find these tools in the Custom Tools menu. Model pressing the **Custom** tool icon and choosing the **Fourths** tool. Move your pointer over the point labeled 0. Make sure the point is highlighted and then click. Move the pointer. As you do, you'll see that you're dragging a "stretchy fraction tool." Its length changes but is always divided into four equal parts by tick marks. Anchor the other endpoint of your stretchy fraction tool by dragging to the point at 1 and clicking when you see that the point is highlighted. Students should note that using the **Fourths** tool has divided the interval from 0 to 1 into four equal parts.

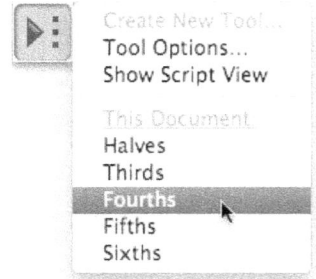

   > Create New Tool...
   > Tool Options...
   > Show Script View
   >
   > This Document
   > Halves
   > Thirds
   > **Fourths**
   > Fifths
   > Sixths

Explain that Sketchpad uses the form 1/4 to write $\frac{1}{4}$.

2. When students have identified the location of $\frac{1}{4}$, model labeling the point. Using the **Text** tool, double-click the point. Type 1/4 and click **OK.** If necessary, drag the label using the **Text** tool, so that the label is below the point.

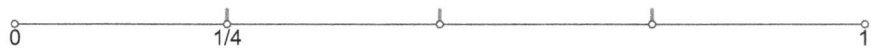

   ```
   o———————————|———————————|———————————|———————————o
   0          1/4                                  1
   ```

   Ask students to name the locations shown by the other two tick marks. They should note that the location of $\frac{2}{4}$ is also the location of $\frac{1}{2}$.

3. *Suppose we didn't have the* **Fourths** *tool.* Press and hold the **Custom** tool icon to show the list of tools. (You might write the list of stretchy fraction tools on the board.) *Is there another way we could divide the interval from 0 to 1 to find $\frac{1}{4}$? We can use the same tool more than once, or we can use a combination of tools.*

You can also label the point at $\frac{1}{2}$ to help students keep track of the divisions.

Students may suggest using the **Halves** tool to divide the whole (the interval from 0 to 1) in half, and then using the **Halves** tool again to divide the interval from 0 to $\frac{1}{2}$ in half. If this is not suggested, hint,

Suppose we use the **Halves** *tool. What can we do with it?* When the class is ready, choose the **Halves** tool. On the second number line, divide the interval from 0 to 1 in half. Then, using the **Halves** tool again, click the points at 0 and $\frac{1}{2}$.

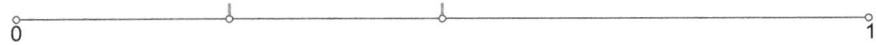

0 ————————|————————|———————————— 1

4. Now that you've found the location of $\frac{1}{4}$ in two different ways, use the **Arrow** tool to drag one number line onto or near the other to help compare the results.

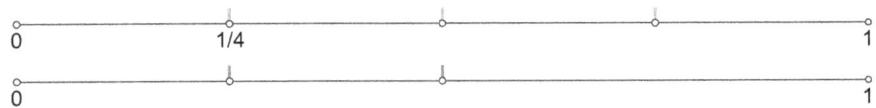

0 ———————1/4———————|————————————— 1
0 ————————|————————|————————————— 1

5. Demonstrate that students can choose **Edit | Undo** to back up one or more steps if they change their minds about what they want to do.

6. If you want students to save their Sketchpad work, demonstrate choosing **File | Save As,** and let students know where to save.

DEVELOP

Expect students at computers to spend about 35 minutes.

7. Assign students to computers and tell them where to locate **Dividing and Subdividing.gsp.** Distribute the worksheet. Tell students to work through step 2 and do the Explore More tasks if they have time.

In the worksheet table, it is important for students to keep track of the custom tools they use and the order in which they use them. Explain to students that they should complete the table carefully so that they can refer back to it later. Encourage students to find and record several ways to locate each fraction.

8. Let students work at their own pace. As you circulate, here are some things to notice.

   • As students create divisions and subdivisions of the 0–1 interval, they should label as many points as they find useful.

   • In worksheet step 2b, students will likely use the **Thirds** tool. To find $\frac{2}{3}$ in a different way, students might experiment and notice that the **Sixths** tool divides the interval into twice as many parts as does the

Thirds tool, with some points in common. In particular, $\frac{2}{3}$ and $\frac{4}{6}$ have the same location.

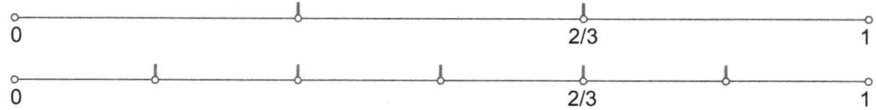

As students solve more problems, they will begin to see that it is helpful to consider the factors of the denominator when looking for different ways to locate the fractions.

- In worksheet step 2c, some students may struggle to locate $\frac{1}{6}$ in a different way after using the **Sixths** tool to do it. Hint, *The whole is divided equally into six parts. It's also divided evenly into other fractional parts. Can you see those? What are those parts?* [Halves and thirds] *How can you use that information to locate $\frac{1}{6}$ with different tools?* Let students investigate this. [The **Halves** and **Thirds** tools can be used together, in either order.]

- Students who are looking for additional ways to find a fraction can change the order in which they use the tools. In worksheet step 2d, for example, students can locate $\frac{3}{8}$ by dividing the interval into fourths and dividing each fourth in half, or by dividing the interval into halves and then dividing each half into four equal parts.

If students have been introduced to reducing fractions, note whether they make the connection between this use of the tools and that procedure.

Be alert to students' thinking about benchmark fractions.

- In worksheet steps 2f and 2i, students may use their knowledge of equivalent fractions to find the fractions using only one tool one time. For students who need a hint, suggest, *There's a way to locate $\frac{8}{10}$ using one tool one time.* Using the **Fifths** tool, students might construct $\frac{4}{5}$ to locate $\frac{8}{10}$; likewise, they might construct $\frac{2}{5}$ to locate $\frac{6}{15}$.

- Students may realize another way to be economical in their use of the fraction tools. In worksheet step 2e, for example, students may begin by dividing the whole interval in half. Because the fraction $\frac{7}{10}$ is larger than $\frac{1}{2}$, students may then divide only the interval from $\frac{1}{2}$ to 1 into tenths, using the **Fifths** tool.

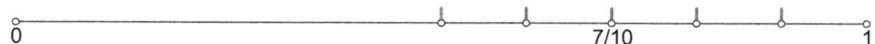

9. If you want students to save their Sketchpad work, demonstrate choosing **File | Save As,** and let them know how to name and where to save their files.

Dividing and Subdividing: Fractions on the Number Line
continued

For **GSP5**   ACTIVITY NOTES

SUMMARIZE

Project the sketch. Expect to spend about 45 minutes.

It's not necessary that students learn the term *lowest terms* or a procedure for reducing fractions at this point. Keep the focus on equivalency.

10. Gather the class. Students should have their worksheets with them. Focus the discussion on the ideas here that are most appropriate for your students.

- If students have previously made paper fraction strips by folding, ask them how subdividing the interval from 0 to 1 using the stretchy fraction tools compares to folding.

- Explore fractions that are not in lowest terms; these are opportunities for students to develop their understanding of equivalent fractions. Invite students who used the **Fifths** tool to locate $\frac{8}{10}$ and $\frac{6}{15}$ to explain their solutions. Propose that the class find more fractions this way. *I'm wondering about $\frac{6}{8}$. Can we find it using one tool one time?* Students will likely see that the **Fourths** tool can be used because the location of $\frac{3}{4}$ is also the location of $\frac{6}{8}$. Challenge the class to locate $\frac{10}{20}, \frac{4}{12}, \frac{12}{16}$, and $\frac{50}{60}$ using one tool one time.

- *Sometimes you were able to locate a fraction by using one tool two times. I saw you locate $\frac{1}{9}$ that way. What other fractions can you locate that way?* $\left[\frac{1}{4}, \frac{1}{9}, \frac{1}{16}, \frac{1}{25}, \text{and } \frac{1}{36}\right.$, as well as any other fraction with one of these denominators$\left.\right]$ *What do you notice about the denominators in all of these fractions?* [They are square numbers.] *Why does using one tool twice create fractional parts whose denominators are square numbers?* Students may give examples such as this one: *Thirds divided into thirds makes 3 times 3 parts. That's how you get a square number. You multiply a number times itself.*

- *Can you tell by looking at a fraction whether you can make it using one tool? By using two tools? How?*

- *Can you tell whether you have located a fraction in all the ways possible using the fraction tools you have? How?*

- If students worked on the Explore More tasks, discuss them. Alternatively, do one or more of the tasks as a class now.

EXTEND

1. *What questions occurred to you about locating fractions?* Encourage all curiosity. Here are sample student queries.

   Why is $\frac{1}{2}$ of $\frac{1}{3}$ the same as $\frac{1}{3}$ of $\frac{1}{2}$?

   What would happen if we had different stretchy fraction tools available? For instance, what if we had tools for primes only? That would be Halves, Thirds, Fifths, Sevenths, Elevenths, and so on. Would that work, too?

   *When I found $\frac{1}{6}$, I used the **Halves** and **Thirds** tools. The numbers 2 and 3 are both factors of 6. Why does this always work, that you can use the factors? Why do the factors of the denominator show how you can divide parts into parts to locate a fraction?*

2. Use page "Explore More 1" to provide more fractions between intervals other than 0–1 for students to locate. To change the label for an endpoint of an interval, use the **Arrow** tool to double-click the point, type a new label in the dialog box that appears, and click **OK.** To change the fraction students are to locate, edit the caption at the top of the page.

ANSWERS

2. a. Use the **Fourths** tool once, or use the **Halves** tool twice.

   b. Use the **Thirds** tool once, or use the **Sixths** tool once.

   c. Use the **Sixths** tool once; or use both the **Halves** and **Thirds** tools, in either order.

   d. Use the **Halves** tool repeatedly to find eighths; or use both the **Halves** and **Fourths** tools, in either order.

   e. Use both the **Halves** and **Fifths** tools, in either order.

   f. Use both the **Halves** and **Fifths** tools, in either order; or, because $\frac{8}{10}$ is equal to $\frac{4}{5}$, use the **Fifths** tool only.

   g. Use either the **Halves** and **Thirds** tools, the **Thirds** and **Fourths** tools, or the **Halves** and **Sixths** tools.

   h. Many combinations of the **Halves, Thirds, Fourths,** and **Sixths** tools work.

   i. Use both the **Thirds** and **Fifths** tools, in either order; or, because $\frac{6}{15}$ is equal to $\frac{2}{5}$, use the **Fifths** tool only.

      j. Use the **Thirds** tool twice.

      k. Use the **Halves** tool repeatedly or use both the **Halves** and **Fourths** tools.

3. Answers will vary.

4. One way to construct $\frac{4}{3}$ is to use the **Halves** tool to locate 1 and then use the **Thirds** tool to subdivide the interval from 1 to 2.

5. One-seventh is between $\frac{1}{6}$ and $\frac{1}{8}$, both of which can be found using the given custom tools; find both fractions and say that $\frac{1}{7}$ is between them. Alternatively, $\frac{1}{7}$ is equal to $\frac{2}{14}$. A fraction close to $\frac{2}{14}$ is $\frac{2}{15}$, which can be constructed by using the **Thirds** and **Fifths** tools. For an even better approximation, consider that $\frac{1}{7}$ is equal to $\frac{3}{21}$. A fraction close to $\frac{3}{21}$ is $\frac{3}{20}$, which can be constructed by using the **Fourths** and **Fifths** tools.

Dividing and Subdividing

For GSP5   Name:

Locate fractions on a number line.

EXPLORE

1. Open **Dividing and Subdividing.gsp**.
   On each page, "a" through "k," there is a fraction to locate.
   Use one or more stretchy fraction tools to locate the fraction.
   If you need to, choose **Edit | Undo** to back up a step.
   To label a point, double-click it. Type the fraction and press **OK**.

2. There is more than one way to find each fraction. Record ways you find. List the tools you use *in the order you use them*.

	Way 1	**Way 2**	**Way 3**	**Way 4**
a. $\frac{1}{4}$	Fourths	Halves Halves		
b. $\frac{2}{3}$				
c. $\frac{1}{6}$				
d. $\frac{3}{8}$				
e. $\frac{7}{10}$				
f. $\frac{8}{10}$				
g. $\frac{11}{12}$				
h. $\frac{5}{24}$				
i. $\frac{6}{15}$				
j. $\frac{1}{9}$				
k. $\frac{1}{32}$				

New York City Title I Elementary School Activities with The Geometer's Sketchpad
© 2012 Key Curriculum Press

EXPLORE MORE

3. Go to page "Make Your Own."

   Name a fraction. Locate the fraction or ask a partner to locate it.

   Name some fractions you cannot locate using the custom tools you have.

   _____

4. Go to page "Explore More 1."

   Now you will locate a fraction greater than 1.

5. Go to page "Explore More 2."

   Locate a fraction as close to $\frac{1}{7}$ as you can.

Jump Along: Equivalent Fractions on the Number Line

INTRODUCE

Project the sketch for viewing by the class. Expect to spend about 20 minutes.

1. Open **Jump Along Equivalent Fractions.gsp.** Go to page "Jump to 1/2." Explain, *Today you'll be using number lines to explore fractions. You'll direct the rabbit to jump along the number line. There are two ways to control the rabbit: You can change how many jumps it will make, and you can change the size of each jump.*

Notice that the top number line represents 1 as $\frac{2}{2}$.

2. Enlarge the window so that it fills most of the screen. Students will be focusing on the interval from 0 to $\frac{2}{2}$. To make this interval easier to see, drag the point at $\frac{2}{2}$ to the right.

3. Ask students to predict what will happen when you press the *Jump Along* button. Take responses and then press the button. The rabbit takes 1 jump (because *Number of Jumps* is equal to 1) and the size of the jump is $\frac{1}{2}$ (because *Jump By* is equal to $\frac{1}{2}$). The rabbit lands at $\frac{1}{2}$, leaving a trace of its jump.

4. Distribute a copy of the Jump Along table from the worksheet to each student. Explain that students will be finding different ways for the rabbit to reach $\frac{1}{2}$. They should fill in the first row of the table with a "1" in the Number of Jumps column because one jump of $\frac{1}{2}$ landed the rabbit at $\frac{1}{2}$.

Although it is possible to enter a non-integer value for the *Number of Jumps* parameter, students will use only integer values in this activity.

5. Using the **Arrow** tool, select the denominator of the *Jump By* fraction. Press the **+** key on your keyboard once to increase the value of the denominator by one. The value of *Jump By* is now $\frac{1}{3}$. Ask students what happened to the fractions along the top number line. Students should note that the top number line is now divided into thirds. The bottom number line has not changed. For students who need to see the prior number line again, press the **−** key on your keyboard to decrease the denominator by 1.

6. With *Jump By* set at $\frac{1}{3}$, ask students what number of jumps will land the rabbit at $\frac{1}{2}$. By looking at the two number lines, some students may reason that one jump of $\frac{1}{3}$ is not enough, while two jumps of $\frac{1}{3}$ is too much. Other students may say that the number of jumps must be half of 3, but 1.5 is not a whole number and only whole numbers are allowed. Because it is not possible to reach $\frac{1}{2}$ with jumps of size $\frac{1}{3}$, tell students to place an **X** in the Number of Jumps column of the table.

The rabbit always starts its jumps at 0. Pressing *Reset* moves the rabbit back to 0 without erasing the traces.

7. Change *Jump By* to $\frac{1}{4}$, discuss what happened to the markings along the top number line, and ask the class how many jumps are needed to land the rabbit at $\frac{1}{2}$. Before testing to see that two jumps of $\frac{1}{4}$ works, select the point below the rabbit and choose **Display | Color** to pick a different color. Now, when you press *Jump Along,* the trace of the rabbit's jumps will be in the new color.

8. Explain that students will be trying the other *Jump By* numbers in the table to see which ones allow the rabbit to land at $\frac{1}{2}$. **Be on the lookout for patterns! Investigate the traces on screen and the numbers in the table.**

DEVELOP

Expect students at computers to spend about 45 minutes.

9. Assign students to computers and tell them where to locate **Jump Along Equivalent Fractions.gsp.** Distribute the worksheet, including multiple copies of the table. Tell students to work through steps 1–9, using a new copy of the table for each new landing spot. Encourage students to ask a neighbor for help using Sketchpad if they have questions.

10. Let pairs work at their own pace. As you circulate, listen to students' conversations. Here are some things to notice.

   • In worksheet step 2, students may use the two number lines together to tell whether it is possible to reach $\frac{1}{2}$ for a particular *Jump By* number. When *Jump By* equals $\frac{1}{5}$, for example, none of the one-fifth intervals on the top number line coincide with $\frac{1}{2}$ on the bottom line. Thus no number of one-fifth jumps will land the rabbit at $\frac{1}{2}$. Other students may figure out how to land at $\frac{1}{2}$ by finding equivalent fractions such as $\frac{2}{4}, \frac{3}{6}, \frac{4}{8}$, and so on. Knowing that $\frac{1}{2}$ is the same as $\frac{3}{6}$, for example, helps students to see that three jumps of $\frac{1}{6}$ will work.

Jump Along: Equivalent Fractions on the Number Line
continued

It is not possible to erase one trace without erasing all the traces.

- In worksheet step 2, the rabbit traces out ways to get to $\frac{1}{2}$. Ask students what they notice. The number of jumps needed to reach $\frac{1}{2}$ increases by one for each trip in the table. When the rabbit jumps twice to reach $\frac{1}{2}$, it divides the interval from 0 to $\frac{1}{2}$ into two equal parts. When the rabbit jumps three times to reach $\frac{1}{2}$, it divides the same interval into three equal parts, and so on.

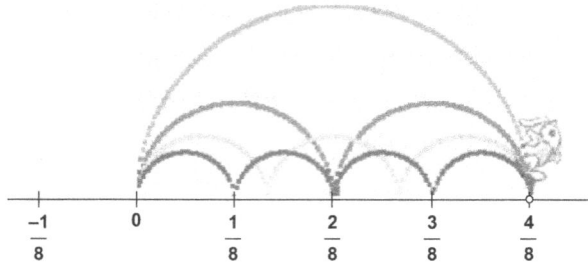

This is a good place to stop on the first day.

- In worksheet step 3, ask students what patterns they notice in their tables. Students may say that they can't make the rabbit jump to $\frac{1}{2}$ when the denominator of the *Jump By* fraction is an odd number. They may also say that the *Number of Jumps* value keeps increasing by one as they look down the column in the table.

There are many ways to represent $\frac{1}{4}$ as a fraction. Equivalent fractions such as $\frac{2}{8}$ and $\frac{3}{12}$ will also work.

11. Introduce worksheet steps 10–12. Now students will keep the number of jumps the same (2 jumps) and enter different *Jump By* values that land the rabbit at $\frac{1}{2}$, at 1, and at $\frac{4}{3}$.

12. As you circulate, listen to students' conversations. Some students might say, *We think there's only one way to reach $\frac{1}{2}$. Take 2 jumps of $\frac{1}{4}$.* These students are correct in thinking that only a jump whose size is $\frac{1}{4}$ will work. Extend their thinking by asking, **Is there another way to tell the rabbit to take a jump of $\frac{1}{4}$?** Let students think about this question and talk with neighbors if they need to.

SUMMARIZE

Project the sketch. Expect to spend about 25 minutes.

13. Gather the class. Students should have their worksheets with them. Direct students' attention to the table they completed in worksheet step 2. **You found that 2 jumps of $\frac{1}{4}$ landed the rabbit at $\frac{1}{2}$. Can we write this information as a number sentence?** Work with students to develop the number sentences $\frac{1}{4} + \frac{1}{4} = \frac{1}{2}$ and $\frac{1}{2} = \frac{1}{4} + \frac{1}{4}$. Some students may also suggest $2 \times \frac{1}{4} = \frac{1}{2}$.

14. *The name of this activity is Jump Along: Equivalent Fractions on the Number Line. Why do you think the title mentions equivalent fractions?* Take responses. Students may say that there are lots of ways to reach a fraction like $\frac{1}{2}$; other fractions are also names for the same location on the number line. Some of these are $\frac{2}{4}, \frac{3}{6}, \frac{4}{8}, \frac{5}{10}$, and $\frac{6}{12}$. As a different example, each table in worksheet steps 10–12 contains a list of equivalent fractions.

EXTEND

What other questions can you ask about jumping along the number line? Encourage student curiosity. Here are some sample student queries.

In how many ways can we reach $\frac{6}{8}$ if we make the denominator of Jump By *equal to 8?*

In how many ways can we reach $\frac{12}{15}$ if we make the denominator of Jump By *equal to 15?*

Are there locations along the number line we can't reach using the Jump By *fractions in the* Jump Along *table?*

How close can we get to $\frac{1}{2}$ without exactly reaching it? How close can we get to 1 without exactly reaching it?

What happens if we make the Jump By *number negative?*

ANSWERS

2. 1 jump of $\frac{1}{2}$, 2 jumps of $\frac{1}{4}$, 3 jumps of $\frac{1}{6}$, 4 jumps of $\frac{1}{8}$, 5 jumps of $\frac{1}{10}$, and 6 jumps of $\frac{1}{12}$

3. The number of jumps needed to land the rabbit at $\frac{1}{2}$ increases by 1, from 1 to 6. Only *Jump By* fractions with even denominators allow the rabbit to land at $\frac{1}{2}$. The number of jumps is always half the denominator of the *Jump By* fraction.

4. 3 jumps of $\frac{1}{4}$, 6 jumps of $\frac{1}{8}$, and 9 jumps of $\frac{1}{12}$

5. The number of jumps needed to land the rabbit at $\frac{3}{4}$ is always a multiple of 3. The denominators of the *Jump By* fractions are all multiples of 4.

6. 2 jumps of $\frac{1}{2}$, 3 jumps of $\frac{1}{3}$, 4 jumps of $\frac{1}{4}$, 5 jumps of $\frac{1}{5}$, and so on

7. It is possible to land at 1 for any of the *Jump By* fractions. The number of jumps is always equal to the denominator of the *Jump By* fraction.

8. 4 jumps of $\frac{1}{3}$, 8 jumps of $\frac{1}{6}$ and 12 jumps of $\frac{1}{9}$

9. The number of jumps needed to land the rabbit at $\frac{4}{3}$ is always a multiple of 4. The denominators of the *Jump By* fractions are all multiples of 3.

10. *Jump By* values of $\frac{1}{4}$, $\frac{2}{8}$, $\frac{3}{12}$, and $\frac{4}{16}$ all land the rabbit at $\frac{1}{2}$. All of these values are equivalent to $\frac{1}{4}$.

11. *Jump By* values of $\frac{1}{2}$, $\frac{2}{4}$, $\frac{6}{9}$, and $\frac{4}{16}$ all land the rabbit at 1. All of these values are equivalent to $\frac{1}{2}$.

12. *Jump By* values of $\frac{2}{3}$, $\frac{4}{6}$, $\frac{6}{9}$, and $\frac{8}{12}$ all land the rabbit at $\frac{4}{3}$. All of these values are equivalent to $\frac{2}{3}$.

14. Problems will vary.

Jump Along Fractions

For GSP5 Name:

Find ways the rabbit can jump to a number on the number line.

1. Open **Jump Along Equivalent Fractions.gsp**. Go to page "Jump to 1/2."

   A rabbit is set to jump along the number line.

   To change the size of the jumps and the number of jumps, select a number and press + or − on your keyboard.

2. Use a Jump Along table. For each *Jump By* number, try to land the rabbit on $\frac{1}{2}$.

   If the rabbit lands on $\frac{1}{2}$, record the number of jumps.

   If the rabbit can't land on $\frac{1}{2}$, put an X.

   Look for patterns that will help you get the rabbit to $\frac{1}{2}$.

   Make each set of jumps a different color. Select the point below the rabbit and choose a new color using **Display| Color.**

3. Look at your table and the jumps traced by the rabbit.
   What patterns do you see?

   _____

   _____

   _____

4. Go to page "Jump to 3/4." Now, the rabbit must land on $\frac{3}{4}$. Use a new table. Try all of the *Jump By* numbers. Record your findings.

5. Look at your table and the jumps traced by the rabbit.
   What patterns do you see?

   _____

   _____

   _____

6. Go to page "Jump to 1." Now the rabbit must reach 1.
   Record your findings in a new table.

7. Look at your table and the jumps traced by the rabbit.
   What patterns do you see?

   _____

   _____

   _____

8. Go to page "Jump to 4/3." Now, the rabbit must reach $\frac{4}{3}$.
   Record your findings in a new table.

9. Look at your table and the jumps traced by the rabbit.
   What patterns do you see?

   _____

   _____

   _____

In steps 10, 11, and 12, the rabbit must take exactly two jumps.

10. Go back to page "Jump to 1/2." Find four ways the rabbit can land
    on $\frac{1}{2}$. Record them in this table.

    Land on 1/2

Jump By	Number of Jumps
	2
	2
	2
	2

New York City Title I Elementary School Activities with The Geometer's Sketchpad
© 2012 Key Curriculum Press

11. Go to page "Jump to 1." Find four ways the rabbit can land on 1.

Land on 1

Jump By	Number of Jumps
	2
	2
	2
	2

12. Go to page "Jump to 4/3." Find four ways to land on $\frac{4}{3}$.

Land on 4/3

Jump By	Number of Jumps
	2
	2
	2
	2

EXPLORE MORE

13. Go to page "Explore More." Make a jump-along problem.

    Choose a place on the number line to land on and name it as a fraction. Decide how many jumps and the jump size.

    For practice, pretend you'd like to jump to $\frac{3}{5}$. Change the *Number of Divisions* to 5. Now the bottom number line will be divided into five equal parts between 0 and 1.

    A Make a caption "3/5" and drag it below the point at $\frac{3}{5}$. Use this point to help you figure out ways the rabbit can jump to $\frac{3}{5}$.

14. On the back of this sheet, record your jump-along problem and ways you solved it.

JUMP ALONG

Land on _____

Jump By	Number of Jumps
$\frac{1}{2}$	
$\frac{1}{3}$	
$\frac{1}{4}$	
$\frac{1}{5}$	
$\frac{1}{6}$	
$\frac{1}{7}$	
$\frac{1}{8}$	
$\frac{1}{9}$	
$\frac{1}{10}$	
$\frac{1}{11}$	
$\frac{1}{12}$	

Land on _____

Jump By	Number of Jumps
$\frac{1}{2}$	
$\frac{1}{3}$	
$\frac{1}{4}$	
$\frac{1}{5}$	
$\frac{1}{6}$	
$\frac{1}{7}$	
$\frac{1}{8}$	
$\frac{1}{9}$	
$\frac{1}{10}$	
$\frac{1}{11}$	
$\frac{1}{12}$	

New York City Title I Elementary School Activities with The Geometer's Sketchpad
© 2012 Key Curriculum Press

Missing Pieces:
Polygons That Keep Their Perimeter

INTRODUCE

Project the sketch for viewing by the class. Expect to spend about 10 minutes.

1. Open **Missing Pieces.gsp.** Enlarge the window so it fills most of the screen. Go to page "One Tile." Scroll down the page if necessary to show all the polygons and the second discard pile. Explain that each polygon is composed of square tiles measuring 1 unit on a side and 1 square unit in area. Check that students can find the perimeter and area of the blue rectangle. [12 units and 8 square units]

2. Ask, ***What do you think will happen to the perimeter of the blue rectangle if I remove one tile?*** Entertain responses. Drag the tile from the location shown at right to a discard pile.

   Is the perimeter still 12 units? [No.] This result will probably not surprise students: A common student misconception is that when the area of a polygon changes, the perimeter (the distance around) also changes.

   Check students' understanding of the term *perimeter*. They should know that it refers both to the boundary around a two-dimensional shape and to the distance around it. Tell students that in this activity, talking about a change in the perimeter refers to a change in the distance around.

Tiles that touch only at a vertex are not allowed.

3. Explain that students will use this Sketchpad model on their own. ***Look for ways to remove one or more tiles from a polygon without changing the perimeter.*** When students remove tiles, all remaining tiles must have at least one side touching the side of another tile; that is, the figures must be polygons. If students cannot find a way to keep a polygon's perimeter the same, they should drag the entire polygon to one of the discard piles. Model using a selection rectangle to select a polygon, and then drag the polygon to a discard pile.

   Students may wonder about removing tiles from the interior of a polygon, leaving a hole. Tell them that this will not be considered in this activity.

4. If you want students to save their work, demonstrate choosing **File | Save As,** and let them know how to name and where to save their files.

DEVELOP

Expect students at computers to spend about 35 minutes.

5. Distribute the worksheet. Assign students to computers and tell them where to locate **Missing Pieces.gsp.** Students should follow steps 1–3 of the worksheet and do the Explore More tasks if they have time. Encourage students to ask a neighbor for help if they have questions about using Sketchpad.

6. Let students work at their own pace. As you circulate, observe and pose questions to learn about students' thinking. What follows are examples of student thinking, suggestions for questions to ask students, and questions to ask yourself.

   • At the start, students may find the perimeter of a polygon by counting units or adding the lengths of all sides, then remove a tile, and find the new perimeter by counting units or adding the lengths of all sides again. At right, students might compute 2 + 4 + 2 + 4 to start, and compute 2 + 4 + 2 + 2 + 1 + 1 + 1 + 1 after removing a tile.

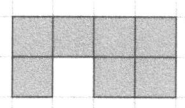

   • Notice when students realize that it is not necessary to compute the entire perimeter. You might prompt, ***Can you find a faster way to check whether the perimeter changes when you remove a tile?*** Here are several methods.

   a. Removing a non-corner tile affects the length of one side only. In the following illustration, the length of the bottom side of the rectangle is 4 units, to start. Removing a tile creates 3 units where there had been 1 unit. Thus, the perimeter of the rectangle changes by an additional 2 units.

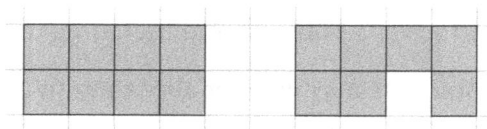

   b. A corner tile contributes 2 units to the perimeter. When the tile is removed, the rectangle loses those 2 units, but gains 2 units from the newly exposed tiles. The perimeter doesn't change.

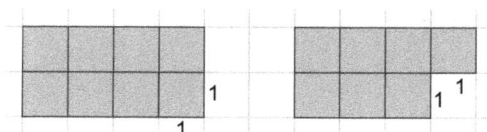

c. The blue rectangle begins with 4 units along its bottom side. When a corner tile is removed, as shown at right in the following illustration, the new polygon still has 4 units "along the bottom," but one of those units has shifted up. The student sees 4 units on each long side of the rectangle, and 2 units on each short side—the same as at the start.

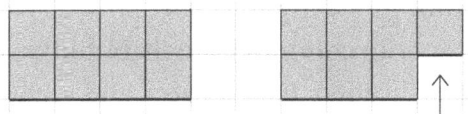

- Some students may think the solutions shown below are different. Clarify that they should be considered the same. Can students see that both solutions involve removing one tile from a corner? Do they see that the polygons are congruent?

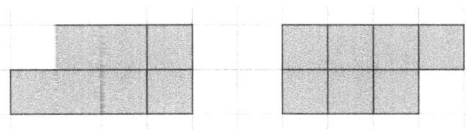

- On page "Two Tiles," do students try to apply what they've learned about removing one tile? In the example that follows, a student began with the 4-by-4 polygon and removed a corner tile to keep the perimeter the same. She then looked for another corner tile to remove. All of the tiles shown with dots are corner tiles (tiles with exactly two sides exposed). Removing any of them will not change the perimeter.

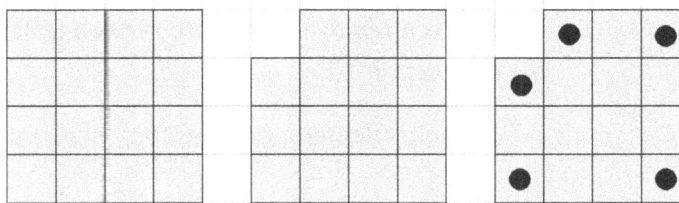

- To challenge students working on the Explore More tasks, you might ask, *What is the largest number of squares you can remove and keep the perimeter the same?*

7. If students will save their work, have them do so now.

SUMMARIZE

Project the sketch. Expect
to spend about 15 minutes.

8. Gather the class. Ask volunteers to demonstrate how they removed one or more tiles to keep the perimeters of polygons the same. Students should discuss polygons for which a solution was not possible.

9. ***Do you have a rule that tells when a tile can be removed without changing the perimeter?*** Provide plenty of opportunity for students to try to communicate their thinking clearly. Encourage debate, clarification, justification, and verification. [Any tile that has exactly two sides exposed can be removed without changing the perimeter.]

EXTEND

What other questions occur to you about the area and perimeter of polygons? Encourage curiosity. Here are sample student queries.

Can you add square tiles to a polygon without changing its perimeter?

Can you add or remove squares from a polygon without changing its area?

What's the most the perimeter of a rectangle can grow if one tile is removed? Two tiles? Three tiles?

What's the greatest number of tiles we can remove without changing the perimeter?

Is there a polygon that you can remove three tiles from, but not four tiles, without changing the perimeter?

What if you remove a tile from the inside of the polygon? Does that change the perimeter?

What about tiles with other shapes, like hexagons?

ANSWERS

2. Only tiles with exactly two sides exposed can be removed without changing the perimeter of one of the polygons. There is no way to keep the perimeter of a $1 \times n$ rectangle the same, because there are no corner tiles.

3. Again, each tile removed must have exactly two sides exposed. The illustration shows two ways to remove two tiles from the blue rectangle without changing its perimeter. On the left, removing the tile in position 1 makes the tile in position 2 a candidate for removal: it now has two sides exposed. On the right, the tiles removed each had exactly two exposed sides at the start.

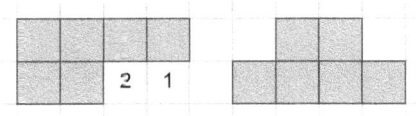

For the red square, two tiles can be removed following the "exactly two exposed sides" rule, but the resulting figure is not a polygon.

4. Any tiles with exactly two sides exposed, originally or along the way, can be removed. The illustration here shows two ways to remove three tiles, using the blue rectangle as an example. As each tile is removed, new sides are exposed, making other tiles candidates for removal.

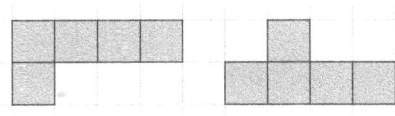

5. Two possible solutions are presented here; others are possible. When a tile with one vertex exposed is removed, the perimeter decreases by 1 unit. When a tile with one side exposed is removed, the perimeter increases by 1 unit. For each vertex-exposed tile removed, a one-side-exposed tile has also been removed in order to keep the perimeter the same.

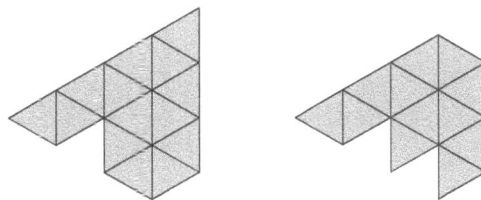

Missing Pieces

Name:

Explore the area and perimeter of polygons.

EXPLORE

1. Open **Missing Pieces.gsp.** Go to page "One Tile."

   Enlarge the window so it fills most of your screen.

   To remove a tile, drag it to one of the discard piles.

   To move a polygon to a discard pile, first select the polygon with a selection rectangle and then drag.

2. Try to remove one tile from each polygon without changing its perimeter.

   What did you discover? What can you say about the tiles that can be removed?

3. Go to page "Two Tiles." Try to remove two tiles from each rectangle without changing its perimeter.

   What did you discover? What can you say about the tiles that can be removed?

EXPLORE MORE

4. Go to page "Explore More 1." Try to remove three or more tiles from each polygon without changing its perimeter.

   What did you discover? What can you say about the tiles that can be removed?

5. Go to page "Explore More 2." Try to remove one or more tiles from each big triangle without changing the perimeter.

   What did you discover? What can you say about the tiles that can be removed?

Running Around the Park: Introducing Perimeter

INTRODUCE

Project the sketch on a large-screen display for viewing by the class. Expect to spend about 10 minutes.

1. Open **Running Around the Park.gsp** and go to page "Runners." Introduce the model. *Who can tell us what a relay runner is?* Take responses. *In this model, relay runners, represented by colored circles, are positioned at the six corners of a park.*

2. In order, press the *Run!* buttons. Each runner should start where the previous runner stopped. Ask students to share their observations. *What happened?* Highlight the language students use to talk about the runners' course. Students might use the words *path, edge, boundary, outside,* and *perimeter.*

Students may also think of the word *periscope. Scope* comes from a Greek word meaning "to watch, or look at." A periscope is a device that lets the user look around.

Introduce *perimeter* if students don't. Explain that *perimeter* is the term used for the distance around a closed (plane) figure or region. Explain the etymology of the word. It is Greek in origin and comes from two words, *peri* (meaning "around") and *metron* (meaning "measure"). *So, perimeter means "the measure around."* Point out the grid the park diagram is on. Explain that each side of a grid square stands for 1 kilometer.

The word **perimeter** *is also used for the boundary around a closed (plane) figure or region.* Explain that in this activity the class will use the word perimeter to refer both to the boundary of the park and to the distance around the park.

3. Explain that students will first find the perimeter of some parks and then make parks themselves. Go to page "Make a Park 1." Demonstrate that the shape and side lengths of this park can be changed by dragging any of the labeled points. Explain the rules for making parks: Each side of a park must lie along a grid line, and the sides of the park must meet at right angles.

4. If you want students to save their Sketchpad work, demonstrate choosing **File | Save As,** and let them know where to save.

DEVELOP

Expect students at computers to spend 35 minutes.

5. Assign students to computers and let them know where to locate **Running Around the Park.gsp.** Distribute the worksheet. Tell students to work through steps 1–9.

Students may not know whether to count the distance at corners as 2 kilometers because two runners each ran a kilometer, or 1 kilometer because the two runners ran around the same square. Choose **Graph | Hide Grid** and give students the opportunity to consider the perimeter again.

6. As you circulate, here are some things to notice about students' approaches to the perimeter problems on page "Runners."

   • Examine students' recordings and observe their ways of finding the perimeters of the parks.

   • In worksheet steps 4 and 6, ask, *How did you figure out the perimeter of the park?* Did students add up the distances in order? Did they add "nice numbers" to make it easier? Did they find the sum of two distances and double it?

   • Some students may have trouble answering the question in steps 4 and 6 because they have not made the connection between the total distance traveled by the runners and the perimeter of the park. Pose questions to help them understand why these values are the same.

7. In worksheet steps 7–9, observe the strategies students use to make parks.

   • In step 7, notice students who create a 2×6 rectangle. Ask, *Why is the perimeter 12 kilometers?* If they count the squares within their park boundary, first watch to see whether they catch their error. If not, ask, *Were you finding the perimeter or the area when you counted the squares?*

   • Some students may calculate the perimeter of the current park, adjust it to make it either smaller or larger, and then check again to see whether their new perimeter is closer to the target perimeter.

   • Students may turn parks into rectangles, dragging the points so that the park has only four distinct sides. Though this is a valid solution, encourage students to look for solutions in which the park maintains six sides.

8. In worksheet step 10, students make their own park. Remind them to follow the rules they have been using. Ask questions to learn about students' thinking. *How did you choose that perimeter for a park? Are you sure it's possible to make a park with that perimeter? Would a six-sided park with a perimeter of 25 kilometers be possible?*

9. Have students save their work, if desired. If a printer is available, you might have students print some of their parks to share with the class. (Students can pencil in the lengths of each side of the park after printing.) Give these instructions for printing.

    • Choose **File | Print Preview.**

    • In the dialog box that appears, set the image to fit on one page and then click **Print.**

SUMMARIZE

Project the sketch. Expect to spend about 15 minutes.

10. Gather the class. If students have printed some of the parks they made, post them grouped by their perimeters. Alternatively, invite students to the computer to make parks on each of the "Make a Park" pages and the "Your Park" page. Students will see that there is more than one way to make a park with a specified perimeter. As a class, discuss the strategies that students used to make the parks.

11. Give students grid paper and have them work individually to draw parks with other perimeters you specify.

EXTEND

1. Specify a perimeter and ask one or both of these questions.

    Suppose the park could have any number of sides. What could your park look like?

    Suppose the sides of the park did not have to meet at right angles. What could your park look like?

2. *What questions can you pose about park perimeter problems?* Encourage all student inquiry. Mathematical questions of interest include these.

    • How many different parks could there be with a given perimeter?

    • Are perimeters always even numbers?

    • What if the sides could be curved, like real parks?

    • What if you had more than six runners?

    • Do parks with the same perimeter have the same area?

ANSWERS

3. Distances will vary, depending on the problem the computer generates.

4. The perimeter will vary, depending on the problem the computer generates.

5. Distances will vary, depending on the problem the computer generates.

6. The perimeter will vary, depending on the problem the computer generates.

7–10. Answers will vary.

Around the Park

Name:

Learn to make parks with different perimeters.

EXPLORE

1. Open **Running Around the Park.gsp.** Go to page "Runners."

2. Press the *Run!* buttons to send the runners around the park.

3. Each side of a grid square stands for 1 km.
   Record how many kilometers each runner goes.

Orange:	Red:	Green:
Blue:	Purple:	Brown:

4. What is the perimeter of the park? _____

5. Press *New Problem* to make a new park.
   This time, press *Run All the Way Around.* Record how far each runner goes.

Orange:	Red:	Green:
Blue:	Purple:	Brown:

6. What is the perimeter of the park? _____

7. Go to page "Make a Park 1."
   Each side of a grid square stands for 1 km.
   Drag the points to make a park with a perimeter of 12 km.

8. Go to page "Make a Park 2." Make a park with perimeter 18 km.

9. Go to page "Make a Park 3." Make a park with perimeter 24 km.

EXPLORE MORE

10. Go to page "Your Park." Make your own problem.
    Follow the same rules for making parks.

New York City Title I Elementary School Activities with The Geometer's Sketchpad
© 2012 Key Curriculum Press

Grade 4 Activities

Balloon Flight: Understanding Decimal Numbers

INTRODUCE

Project the sketch for viewing by the class. Expect to spend about 10 minutes.

You might relate the landing pad to a number line, where only the part that is the pad is shown.

If students propose testing landing-spot numbers that extend to the hundredths place, ask them to save this idea for later.

1. Display **Balloon Flight.gsp.** Go to page "Pad 1." Explain, *You are going to fly a virtual hot-air balloon. To land the balloon safely, you'll use what you know about decimals.*

2. Draw students' attention to the landing pad and the numbers that mark its endpoints, 4.1 and 4.5. *To land the balloon successfully, you must set its landing spot at a number that is in between 4.1 and 4.5.* Point out the current landing-spot value on the page, 0.0. *You can specify a number that extends to the tenths place.*

3. As students suggest landing-spot values, invite the class to agree with or contest the proposals and explain their thinking. Record all suggestions.

4. Test the suggested values, one at a time, by changing the landing-spot value and pressing *Lift Off!*. (If the value you are testing is far away from the numbers on the landing pad, the balloon may fly off the screen and hang in the air.) Press *Reset* each time and enter a new number. Note which numbers result in a successful landing and which don't. Ask these questions.

   What do you notice about all the numbers for the successful landings?

   Which number lands the balloon closest to the left side of the pad? To the right side? Is it possible to get even closer? Does one number land it in the middle? [4.3]

To "hit the middle" of the landing pad for endpoints such as 5.0 and 5.1, let students discover that they will need precision to the hundredths place.

5. Provide more problems as needed by changing the values of the endpoints to pairs of numbers such as those given here. Avoid pairs whose numbers differ by a tenth. That will be the focus of the next step.

   3.2 and 3.8   10.4 and 10.6   5.0 and 5.7   7.4 and 8.0   0.2 and 0.5

   For each number pair ask, *What number lands the balloon in the middle of the pad?*

DEVELOP

Continue to project the sketch. Expect to spend about 20 minutes.

6. Change the endpoints of the landing pad to 4.0 and 4.1. *Can you name a landing spot that will work?* The precision of the landing-spot value extends only to the tenths place. The class should propose the need for numbers that extend to the hundredths place.

7. Go to page "Pad 2." Again, have students propose landing-spot values, and invite them to agree with or challenge proposals and explain their thinking. Record all proposals, run the balloon flights to test them, and invite discussion of why the landings are or are not successful.

8. If you wish to provide more problems, change the values of the endpoints to pairs of numbers such as these. Challenge students to hit the middle of the pad.

   6.2 and 6.3   10.5 and 10.6   3.9 and 4.0   0.4 and 0.5   7.6 and 7.7

   A common student misconception is thinking that 6.25, for example, is greater than 6.3 "because 25 is more than 3". The following questions may help students examine their thinking.

   What is the value of each of the digits in 0.25?

   Is 0.3 the same as 0.30? Which is greater: 0.30 or 0.25?

9. Draw a landing pad on the board with endpoints 4.00 and 4.01. *What's a number that will allow you to land successfully?*

10. When students have proposed the need for numbers that extend to the thousandths place, go to page "Pad 3."

11. Have students suggest landing-spot values. Run flights to test them. By now, students may be developing generalizations about landing spots that work.

SUMMARIZE

Expect to spend
15 minutes.

12. *I've got some decimal challenges for you. Use what you know about place value and decimals to think about these questions.*

13. Write the numbers 7.39 and 7.1239 on the board. *Which is greater?* Accept responses and encourage the class to verify that each is correct.

    For students who identify 7.1239 as greater because it has more digits to the right of the decimal or because 1239 is greater than 39, rewrite 7.39 as 7.3900 and discuss place value.

14. Write the numbers 0.999 and 1 on the board. *Name a number that is between these numbers.* Invite discussion. Sample student responses include these.

    I know that 0.999 is smaller than 1. You can't add any tenths, hundredths, or thousandths to 0.999 without making it bigger than 1. So, I added a number to the next place: 0.9991. You can get as big as 0.9999. (You may explain, *The next place is the ten-thousandths place. A ten-thousandth is one-tenth of a thousandth.*)

You know that if you add 1 thousandth to 0.999, you get 1. You have to add something smaller than a thousandth to it. You can add 1 ten-thousandth, 2 ten-thousandths, or as many as you want, up to 9 ten-thousandths.

You can add any numbers you want after the nines because all of that will be smaller than adding another thousandth, and you'd have to add a thousandth to get to 1. So, a number in between could be 0.999987654321.

15. *Suppose I name any two decimals. Do you think it would always be possible to find a decimal that is in between them?* You may wish to have students respond individually to this question in writing. Students are likely to believe that it is possible to find such a decimal. This intuition is correct.

EXTEND

Have students record the endpoints and every landing spot they try, including ones that are incorrect. Viewing students' answers will help you assess their thinking about ordering decimals.

1. For students who would benefit from additional time with decimals, provide an opportunity to use the "On Your Own" pages of the sketch, either alone or in pairs. On these pages, Sketchpad presents a new problem every time *New Challenge* is pressed. The precision of the decimals increases from page to page.

2. Pairs of students may exchange challenges using the models on "Pad 1," "Pad 2," and "Pad 3." One student changes the values of the landing pad's endpoints. The other enters a landing spot and launches the balloon.

Zooming Decimals: Precision and Place Value

INTRODUCE

Project the sketch for viewing by the class. Expect to spend about 20 minutes.

1. Open **Zooming Decimals.gsp.** Go to page "Model 1." Distribute the worksheet.

2. Ask students to describe what they see. The model displays a number line labeled from 0 to 10. A red point sits on the line. Drag the point to show that it can move anywhere on the number line. Then choose **Edit | Undo** one or more times to return the point to its original position.

```
0    1    2    3    4    5    6    7    8    9    10
├────┼────┼────┼────┼────┼────┼──o─┼────┼────┼────┤
```

3. *What can you say about the location of the red point?* Give students time to record their answers alongside "First" on the worksheet. Take responses and record them on the board. Sample responses follow.

   The point is between 6 and 7.

   The point is around $6\frac{1}{2}$.

   The point is around 6.5.

   The point is between 6.5 and 6.75; it's more than halfway to 7, but it's not three-quarters of the way.

4. *The point sits somewhere between 6 and 7. How do you think we can find its location more precisely?* Take responses. Students may suggest dividing the interval between 6 and 7 into more parts.

5. *Let's take a closer look at what's happening between 6 and 7.* Press the first *Zoom* button. A number line will appear directly below the 6–7 interval and slowly expand, as if "zooming in" on this interval.

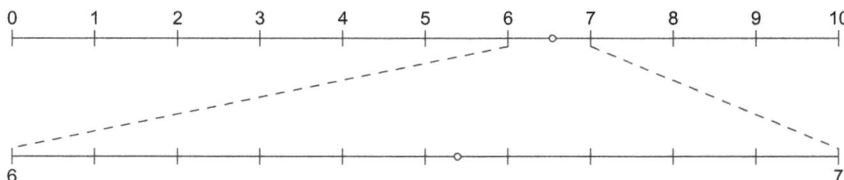

```
0    1    2    3    4    5    6    7    8    9    10
├────┼────┼────┼────┼────┼────┼──o─┼────┼────┼────┤

6                        o                        7
├────┼────┼────┼────┼────┼──o─┼────┼────┼────┼────┤
```

6. *How does this number line relate to the one above it?* Elicit the idea that the new number line represents a magnified, or "zoomed," view of the interval where the point lies, between 6 and 7. The dashed lines connecting the two number lines show which portion of the original number line is shown on the number line below it. Explain that the point sitting on the new number line is "the same" as the one above it:

Both points lie at the same location. This may not be immediately clear to students because the points do not sit one directly below the other.

7. ***What do the tick marks that sit between 6 and 7 represent? How far is it from one tick mark to the next?*** Now, an interval of one has been divided into ten equal parts, so there is an increase of one-tenth, or 0.1, from tick mark to tick mark. Point to each tick mark between 6 and 7, asking the class to count as you go along: *six and one-tenth, six and two-tenths, ...*

8. ***What can you say about the location of the point now?*** Give students time to record their responses alongside "Second" on the worksheet. Take responses and record them on the board. Sample responses follow.

   Now we can estimate the location more accurately.

   We were right that the point is a little closer to 7 than it is to 6.

   The point is between $6\frac{5}{10}$ and $6\frac{6}{10}$.

   The point is between 6.5 (six point five) and 6.6 (six point six).

   The point is around 6.55.

   Give the class time to discuss estimates of the point's location.

9. ***When we zoom in, we gain precision; we can describe the location of the point more accurately. What do you think we'll see if we zoom in again, this time on the interval between 6.5 and 6.6?*** Take responses. Students may or may not predict that the interval will be divided into ten smaller parts, with each part representing a tenth of a tenth—a hundredth.

10. Press the next *Zoom* button. The new interval 6.5 to 6.6 is shown. Elicit the idea that again an interval has been divided into ten equal parts, but this time a tenth has been divided, not one whole unit. ***What is a tenth of a tenth?*** Read the location of each tick mark with the class: *six and fifty-one hundredths, six and fifty-two hundredths,* and so on.

> If you want students to estimate the location of the point to the tenths or hundredths place only, stop here. Press *Reset,* and try a new problem by dragging the point to a new location.

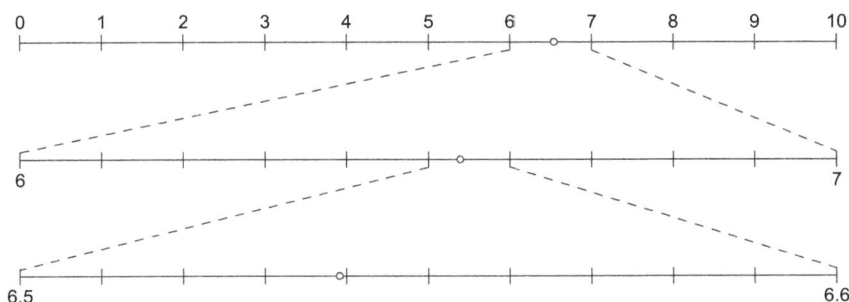

11. Ask students to use this magnified view to make a more precise estimate of the point's location. Students should write their answers on the worksheet alongside "Third." Take responses and record them. Here are samples of student thinking.

    The point is closer to 6.5 than 6.6.

    The point is between six and fifty-three hundredths and six and fifty-four hundredths.

    The point is between 6.53 (six point five three) and 6.54 (six point five four).

    The point is very close to 6.54. I'd say it's probably about 6.539.

12. If you want to continue to thousandths and ten-thousandths, repeat the sequence of steps two more times. Press the next *Zoom* button, watch the interval expand, discuss what the tick marks represent, and ask students to estimate the location of the point. Students' final estimates of the point's location will likely be that it lies between 6.5391 and 6.5392. To view the location of the point, reported to eight decimal places, press *Show Location*.

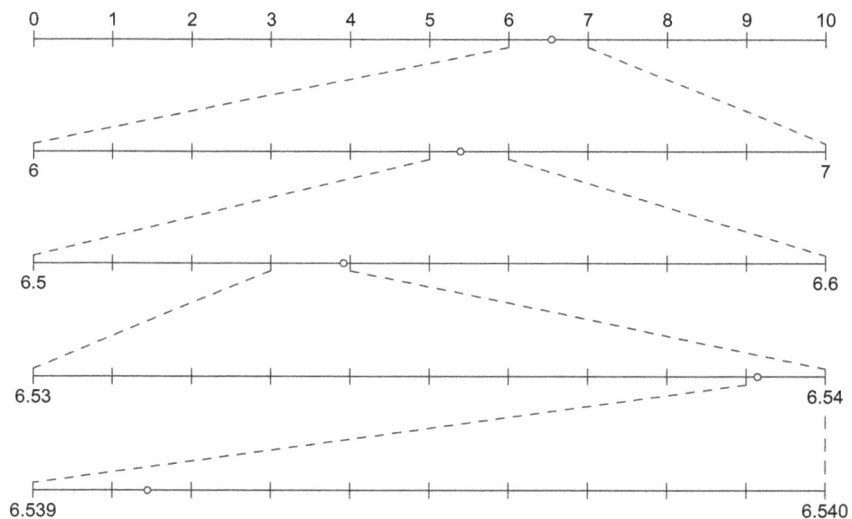

DEVELOP

Continue to project the sketch. Expect to spend about 30 minutes.

13. Have students look at all five estimates on their worksheets. **What is different about the estimates you made using each number line?** Students should explain that each time they viewed a new number line a more detailed scale was shown, allowing them to name the location of the point more precisely.

14. Press *Reset* to hide all but the top 0–10 number line. Drag the point to a new location. Repeat the same steps, having students record their estimates.

15. Press *Show Location*. Ask a volunteer to drag the point that sits on the top number line. Students will observe that all five points move simultaneously, because each point represents the same location.

To speed up the movement of all the points, choose **Display | Show Motion Controller** and click the up arrow repeatedly to increase the speed.

16. To make the movement steadier, press *Animate Point*. **Look at how all five points are moving. What do you notice?** Here are some sample responses.

    The points all move at different speeds.

    The point on the first number line moves the slowest. The point on the fifth number line moves the fastest.

    Every time the point on the fourth number line moves all the way across, the point on the third number line moves one tick mark to the right. That's because the fourth number line is divided into thousandths and the third number line is divided into hundredths. Every time the point has gone 10 thousandths, it has gone a hundredth.

    Every time the point on the fifth number line moves all the way across, the point on the fourth number line moves one tick mark. That's because the fifth number line is divided into ten-thousandths and the fourth number line is divided into thousandths. There are 10 ten-thousandths in a thousandth.

SUMMARIZE

Continue to project the sketch. Expect to spend about 10 minutes.

17. Present problems such as the following. Take responses and have students write each response using decimal notation.

    A point is closer to 3.6 than it is to 3.7. What are some possible locations of the point? Here are two sample responses.

    Three and sixty-four hundredths. [3.64]

    Three and six hundred twenty-five thousandths. [3.625]

    Name a point that is closer to 5.0 than it is to 5.1.

    Name a point that is closer to 99.9 than it is to 100.1.

    Name a point that is closer to 30.45 than it is to 30.47.

18. *Suppose I name two decimals for you. Do you think it's always possible to name another decimal that lies somewhere between them?* Students' experiences with magnifying the number line and viewing ever-finer scales may lead them to believe (correctly) that it is always possible to do so. Provide an opportunity for them to communicate their reasoning and craft explanations in their own words.

EXTEND

The colored diamonds serve as a way to indicate where the hidden point sits along the five number lines.

1. Page "Model 2" contains a related model to explore. Students are given the numerical location of an unseen point and must mark the point's location along the five progressively scaled number lines. To start, students should press *Units* to reveal a number line scaled from 0 to 10. Students should decide on which unit interval the point is located and drag the colored diamond so that it covers that interval. Continuing in this manner with the remaining four number lines, students should identify the intervals on which the hidden point is located and mark them with the colored diamonds. Pressing *Show Answer* reveals the locations of the point. Pressing *New Problem* generates a new number.

2. On page "Model 1," spend time identifying the location of points on intervals other than 0–10. To change the endpoints of the top number line, double-click *Left Endpoint* = 0, enter a whole number in the dialog box, and click OK.

3. For students who would benefit from more individualized work, provide opportunities to use the decimal model alone or in pairs.

4. Discuss the concept of calibrated scales in tools people use. Begin by discussing measuring tools like tape measures and thermometers that students are familiar with. Continue by discussing reasons people use estimates to the tenths, hundredths, thousands, and ten-thousandths places. Include the idea that the "best" estimate is the one with the level of precision needed in a particular situation: *When you bake a cake, do the directions tell you to bake from 30.5 to 39.5 minutes?*

Zooming Decimals

Estimate the location of a point on a number line as you zoom in.

EXPLORE

1. First _____

   Second _____

   Third _____

   Fourth _____

   Fifth _____

2. First _____

   Second _____

   Third _____

   Fourth _____

   Fifth _____

3. First _____

   Second _____

   Third _____

   Fourth _____

   Fifth _____

4. First _____

   Second _____

   Third _____

   Fourth _____

   Fifth _____

5. First _____

   Second _____

   Third _____

   Fourth _____

   Fifth _____

6. First _____

   Second _____

   Third _____

   Fourth _____

   Fifth _____

Identity Properties: Exploring 0 and 1

INTRODUCE

Project the sketch for viewing by the class. Expect to spend about 10 minutes.

1. Open **Identity Properties.gsp.** Go to page "Sum 1." Follow these steps to introduce the model.

   • Explain, *Today we're going to investigate a special code that uses symbols in place of numbers.* Have the class look at the numbers above the symbols and note that the numbers 0 through 9 are shown.

   • Draw attention to the two pointers below the symbols. Ask which numbers the pointers are sitting beneath. [3 and 5]

   • Ask the class to look at the symbol displayed as the sum. *I'm wondering how this code works. Do you have an idea why the sum is the hexagon symbol?* Give students time to think about this question. They should realize that the sum of 3 and 5 is 8, and the hexagon represents 8 in this code.

Some students may think that when the sum shows two symbols side by side, the symbols are being added. Acknowledge this logical interpretation, and then explain that this is not how the code works.

   • Invite a volunteer to the computer to drag one of the pointers. As the pointer moves, the sum will change. For certain positions of the pointers, there will be more than one symbol appearing in the sum. Ask students to think about why this is so. [When the sum of the numbers is equal to or greater than 10, each digit of the sum—the digit in the tens place and the digit in the ones place—is represented by its own symbol.]

   • Draw the symbols ☹ ☐ on the board and ask, *Suppose the sum is frowning face, square. What symbols could the pointers be below?*

   • Give students time to talk in pairs, and then discuss as a class. Take responses until students are satisfied they've named all possible pairs of symbols. The combinations they are likely to suggest are 3 and 9, 4 and 8, and 5 and 7.

   • If students don't suggest placing both pointers below the symbol for 6, or if some students are skeptical about whether the sketch will work in this way, drag both pointers to that position. The sum 6 + 6 is computed.

DEVELOP

Continue to project the sketch. Expect to spend about 30 minutes.

2. Go to page "Sum 2" and introduce this model. (The pointers should be located below 😞 and ☆.) *This new code is like the last one. It has the same symbols. But there are two differences in this model. Can you see one?* [The values of the symbols are not shown.] *The other difference is that the value of each symbol has changed from its value in the previous model.*

   I wonder whether it's possible to tell when one of the two pointers is located below 0. For instance, how about now? Is either frowning face or star equal to 0? Talk to your neighbor and discuss the question.

3. Give pairs a few minutes to think and talk. Then facilitate discussion. The student explanations here represent two ways students frequently think about this problem. Both will be useful as students continue to work.

   Any number plus 0 is that same number. Zero is the only number that doesn't change another number when added to it. If frowning face is 0, then the sum will be star. If star is 0, then the sum will be frowning face.

   Because 0 plus 0 equals 0, we could put both pointers on frowning face to see whether the sum turns out to be frowning face! We could do the same thing with star.

4. *I wonder whether it's possible to figure out which of the ten symbols is equal to 0. Discuss this with your neighbor. Do you think you can find 0? If so, how?*

   Give pairs time to craft their strategies. Then facilitate discussion. Encourage students to ask speakers for clarification as strategies are explained. Provide enough time for most students to understand the two useful strategies explained in the student responses here.

   We'd move the pointers to different locations and check the sums. For each new location, we'd check whether the sum is the same symbol as one of the symbols we're adding. If it is, then we know one of our symbols is 0, and we know which one it is.

   We'd move both pointers below the same symbol. If the sum is equal to that symbol, we're done—that symbol equals 0. If it isn't, we'd just move the pointers below the next symbol. We'd keep doing this until we find 0.

Identity Properties: Exploring 0 and 1
continued

Students can check their work by pressing *Show 0* to reveal 0 and *Show Other Numbers* to reveal 1–9.

5. Invite a volunteer to the computer to try one of the strategies. Explain that the class should monitor how the strategy is being carried out, and offer feedback and suggestions.

6. After the class discovers that ⊓ equals 0, ask students to write on paper or on the board several addition statements that have ⊓ as an addend. For example, ⊓ + ▽ = ▽ or ⊓ + ☆ = ☆. Students should be able to write these statements without manipulating the Sketchpad model.

Exploring the Identity for Multiplication

7. Go to page "Product 1" and introduce this code. (The pointers should be located below □ and ▽.) *Here's a new code. This code is a little different from the previous one. Now, instead of adding symbols, the computer multiplies them.* Make sure students note that the product, not the sum, is displayed and that again the values of the symbols are not shown.

   In the previous model, we used the idea that 0 added to a number doesn't change the number. I wonder whether there is a number that you can multiply other numbers by without changing them? Provide a few minutes for students to talk about this in pairs. Then facilitate discussion. Students should agree that multiplying a number by 1 yields the original number.

8. *I wonder whether we can discover when one of the pointers is located below 1.* (Make sure the pointers are located below □ and ▽.) *For instance, how about now? Is either square or upside-down triangle equal to 1? Talk to your neighbor.* Give students a few minutes to talk. Then facilitate discussion. Here is one typical way that students reason. *Any number times 1 is that same number. No other number works that way. Since upside-down triangle times square equals square, upside-down triangle has to be 1.*

The inquiry is well worth students' time. Note how well students are able to communicate as they explain, challenge, justify, and verify the ideas put forth.

Although it is possible that ▽ equals 1, it is also possible that □ equals 0. If □ does equal 0, then ▽ might not be 1. It's important for students to realize that although they have found one possible explanation, it may not be the only one. Guide them in making this discovery by pushing them to do more thinking. *I'm wondering about this idea. Do you agree that the only way the product can be square is if upside-down triangle is equal to 1?* Provide lots of time for students to tackle this question in pairs and then as a class. The discussion should bring out the role of 0 in multiplication: Any number multiplied by 0 (or 0 multiplied any number of times) yields 0.

9. ***How do you think we can check whether upside-down triangle really does equal 1?*** Have pairs discuss, and then facilitate class discussion. Provide enough time for most students to understand the two strategies described in these student explanations.

   Let's keep one of the pointers on upside-down triangle and move the other pointer from symbol to symbol. On each move, we'll check whether the product is equal to the symbol we moved to. If the product is always equal to the symbol we move to, we know that upside-down triangle is equal to 1.

   Let's move both pointers below upside-down triangle. If the product is upside-down triangle, then we know that upside-down triangle is equal to 1.

   Ask a volunteer to use the Sketchpad model to check whether $\triangledown$ equals 1, and take feedback and suggestions from the class. Let the class discover that $\triangledown$ does indeed equal 1.

10. Distribute paper. Ask students to write several multiplication statements that have $\triangledown$ as a multiplier. For example, $\triangledown \times \sqcap = \sqcap$. Students should be able to write these statements without manipulating the Sketchpad model.

SUMMARIZE

Continue to project the sketch. Expect to spend about 20 minutes.

11. Begin a class discussion by asking, ***What is special about 0 and 1 that helped you to find them in the codes?*** Have students discuss in pairs and then as a class. Here is one possible response: *Both numbers, 0 and 1, didn't change anything. When we added 0 to a number, we got the same number. When we multiplied a number by 1, we got the same number.*

12. ***As you've just explained, 0 and 1 are special. Mathematicians call 0 the identity for addition and 1 the identity for multiplication. Adding 0 to a number doesn't change that number's identity. The same thing is true for multiplying a number by 1. The identity of the number stays the same.***

EXTEND

1. For students who would benefit from more individualized work on finding the symbol for 0 for an addition code and the symbol for 1 for a multiplication code, provide an opportunity to use pages "Sum 3" and "Product 2" of **Identity Properties.gsp** alone or in pairs.

2. Have students discuss how to find the value of 0 in the multiplication code. For your information, on page "Product 1," ☐ is equal to 0 because any number times ☐ equals ☐.

Number Codes:
Properties of Addition and Multiplication

For GSP5

ACTIVITY NOTES

INTRODUCE

Project the sketch on a shared computer for viewing by the class. Expect to spend about 10 minutes.

1. Open **Number Codes.gsp.** Go to page "Sum 1." Explain, *Today we're going to investigate a code that uses symbols in place of numbers.* Use this procedure to acquaint students with the model.

   • Have the class look at the numbers above the symbols and note that every whole number from 0 through 9 appears.

   • Draw attention to the two pointers below the symbols. Ask which numbers the pointers are sitting beneath. (We'll assume here that the pointers are positioned below 3 and 5.)

   • Ask the class to look at the symbol to the right of "SUM = ." *I'm wondering how this code works. Do you have an idea why the sum is* **hexagon?** Give students time to think about this question. They should surmise that because the sum of 3 and 5 is 8, *hexagon* represents 8 in this code.

Some students may think that when the sum shows two symbols side by side, the symbols are being added together. Acknowledge this logical interpretation, but then explain this is not how the code works.

   • Invite a volunteer to the computer to drag one of the pointers. As the pointer moves, the sum will change. For certain positions of the pointers, there will be more than one symbol in the sum. *Why is this so?* Elicit the idea that when the sum of the numbers is a two-digit number, each digit of the sum—the digit in the tens place and the digit in the ones place—is represented by its own symbol. At this point, some students may want to take a minute to figure out what sums are possible when pairs of the whole numbers from 0 to 9 are added. [Zero through 18]

   • *Suppose the sum is* **frowning face, square.** Draw the symbols on the board.

Someone may suggest 10 and 2, but others will probably point out that 10 is not a digit represented by a symbol.

   What symbols could the pointers be under? Give students time to talk in pairs, and then discuss as a class. Invite responses until students are satisfied they have named all possible pairs of symbols. Some combinations are 3 and 9, 4 and 8, and 5 and 7.

   • If students don't suggest placing both pointers below the symbol for 6, drag both pointers to that position. The sum 6 + 6 is computed.

DEVELOP

Continue to project the sketch. Expect to spend about 35 minutes. If time is short, focus only on the addition code.

2. Distribute the worksheet. *Keep track of the number represented by each symbol as we find it.*

3. Go to page "Sum 2." *This new code is like the last one. It has the same symbols, and the symbols represent the numbers 0 through 9. But the symbols may represent **different** numbers from the ones they represented in the last code, and you can't see the numbers! I wonder whether it's possible to find the numbers the symbols represent. What do you think? Talk with your neighbor.*

4. *Do you have any ideas about how to crack this code?* Invite volunteers to the computer to drag the pointers and explain any strategies they have devised. You might also ask a student to drag the pointers around slowly so the class can watch: Do students notice anything that seems interesting?

Some students may want time to figure out what the possible two-digit sums are for any two whole numbers from 0 through 9 to convince themselves that the tens-place digit must be 1.

- Based on the discussion at the beginning of the activity, some students might focus on those sums that have a tens-place digit. If students realize that the tens place can only be 1, they will be able to determine that *star* represents 1.

- Once students have determined the symbol for 1, they may want to look for 2. Because $1 + 1 = 2$, dragging both pointers to sit beneath *star* reveals that *point-down triangle* represents 2.

- If students realize they can continue with this strategy, finding the symbols representing the rest of the numbers through 9 is easy. Dragging the pointers below the symbols for 1 and 2 reveals the symbol for 3. Similarly, pointing to the symbols for 1 and 3 reveals 4, and so on.

- Students may suggest looking for the symbol that represents 0. *How will you look for it? How will you know when you've found it?* Provide time for all students to consider these questions before taking responses. If there are no responses, ask, *What's special about adding 0 to a number?* Students may explain, "Adding 0 to a number doesn't change that number." Invite a volunteer to the computer to drag the

The symbol *placard* looks like a poster on a stick.

pointers until the symbol for 0 is determined. [In this code, adding *placard* to any symbol results in a sum represented by that same symbol. Thus *placard* represents 0.]

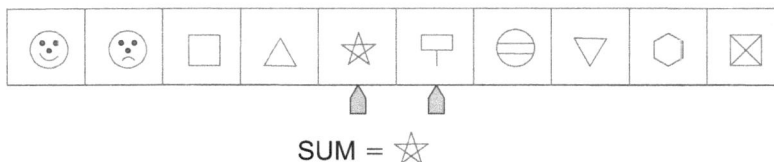

SUM = ☆

5. The page "Sum 3" offers another addition code. Work on it together now or provide it for practice at a later time.

6. Go to page "Product 1." Explain that in this code, the *product* of two symbols is displayed. Again, invite volunteers to the computer to move the pointer in ways the class suggests. Allow the class to experiment and look for patterns as they determine the numbers the symbols represent. Give students the opportunity to develop their own solution strategies. Here are some discoveries they may make and strategies they may devise.

- Any number multiplied by 0 is equal to 0. In this code, any symbol multiplied by *square* yields *square*. Thus *square* represents 0.

- Multiplying a number by 1 does not change its value. Thus in this code, *point-down triangle* represents 1.

- When one pointer is positioned below *smiling face* and the other is dragged across the chart, *square* and *smiling face* alternate in the ones place of the product. What number has multiples in which 0 and another digit alternate in the ones place? The multiples of 5 have a ones-place digit of either 0 or 5. No other number in the code (or in the times table) produces this result. This means that *smiling face* represents 5.

- Place both pointers under the symbol for 5. The symbols for 2 and 5 (for the digits in 25) will appear in the product. The symbol in the tens place represents 2.

- Place both pointers under 2. The result is the symbol that represents 4.

- Place both pointers under 4. The symbol in the ones place of the product represents 6.

- Place both pointers under 6. The symbol in the tens place of the product represents 3.

  - To find 9, multiply 3 × 3. To find 8, multiply 9 × 9. To find 7, multiply 3 × 9.

7. Page "Product 2" offers another multiplication code for students to crack. The class can solve this code now or students can work on it in pairs later.

SUMMARIZE

Working away from the computer if you wish, expect to spend about 15 minutes. If time is short, focus only on the addition code.

8. Close the activity with a discussion of the strategies the class used to crack the codes. ***Suppose we had a visitor walk into our classroom and ask us about this row of symbols. What would be a good way to explain the code and ways to crack it?*** Students might enjoy taking their worksheets home and explaining the code to family members.

9. Highlight the role of the identity element for addition and multiplication. ***When you were solving the codes, you used a special property of the numbers 0 and 1. What is special about those numbers?*** Here are some sample student responses.

   They are the only two numbers that keep the symbols the same when you point to a pair of symbols.

   Zero works for addition, and 1 works for multiplication. They work the same way, but for different operations.

   Knowing this special property of 0 for addition and 1 for multiplication was helpful today when you were breaking the codes. In mathematics, this special property is useful in many other ways you will learn about in the future. In mathematics, we say that 0 is the identity element for addition and 1 is the identity element for multiplication.

EXTEND

1. For students who would benefit from more individualized work on solving number codes, consider giving them opportunities to crack the codes working alone or in pairs.

2. ***What other questions about the number codes can you ask?*** Encourage all student inquiry. Mathematical questions of interest include these.

- What's the best strategy for finding the values of the symbols?

- What's the fewest number of calculations we need to do to determine all the numbers?

- What if we had three markers? Could we find the values with fewer calculations?

- What if our codes showed the difference of two symbols or the quotient of two symbols? How could we solve the codes then?

Crack the Codes

Name:

Find the numbers, 0 through 9, represented by the symbols.
Write the numbers above the symbols.

1. Crack the code on page "Sum 2."

2. Crack the code on page "Sum 3."

3. Crack the code on page "Product 1."

4. Crack the code on page "Product 2."

Jump Along: Multiplication on the Number Line

INTRODUCE

Project the sketch for viewing by the class. Expect to spend about 10 minutes.

1. Open **Jump Along Multiplication.gsp.** Go to page "Jump Along." Ask students to describe what they see. A rabbit is sitting at 0 on a number line.

2. Tell students, *Watch what happens when I press* **Jump Along.** Press *Jump Along.* **What did the rabbit do?** Take responses. If necessary, press *Move Rabbit to 0* and then *Jump Along* so that students can watch the movement again. The rabbit makes four jumps of size 1 and lands at 4.

Although it is possible to enter a non-integer value for the *Number of Jumps* or *Jump By* parameter, students will use only integer values in this activity.

3. Change the *Jump By* number to 2 by double-clicking *Jump By* with the **Arrow** tool, entering 2, and pressing **OK.** Reset the rabbit by pressing *Erase Traces* and *Move Rabbit to 0.* Press *Jump Along.* Ask students to describe what the rabbit did. Make sure everyone sees that *Number of Jumps* tells them how many times the rabbit jumped and *Jump By* tells the size of each jump.

4. Distribute the worksheet. Together, look at the table in step 2. Ask students to fill in all but the last column of row a. Students should relate this row to the rabbit in the sketch: The rabbit made four jumps of size 2 and landed at 8. Point out that the rabbit always starts at 0.

   Let's look at the last column of the table. The rabbit landed at 8. Can you write a number sentence that explains how the rabbit made its way to 8? As a class, develop the number sentence 2 + 2 + 2 + 2 = 8. *Why are we adding 2 each time?* [The size of the rabbit's jump was always 2.] *Why do we add four 2's?* [The rabbit jumped 4 times.]

As students solve more problems, they will make the connection between repeated addition and multiplication.

5. Write 4 on the board and ask, *Suppose I want to write a different number sentence that starts with 4 and means the same thing as 2 + 2 + 2 + 2 = 8. How can we complete this number sentence without using addition signs?* Students may suggest $4 \times 2 = 8$. *How does this multiplication sentence relate to the rabbit's movement along the number line?* [The rabbit made 4 jumps, each of size 2, and landed at 8.]

This is correct because of the *commutative property of multiplication.*

Note that students may also suggest $2 \times 4 = 8$ as a valid multiplication statement. Explain, *That is certainly another correct way to represent 2 + 2 + 2 + 2 = 8. Let's agree as a class to write the number of jumps first.* Establishing this convention with students will facilitate conversations at computers and class discussions.

DEVELOP

Expect students at computers to spend about 30 minutes.

6. Assign students to computers and tell them where to locate **Jump Along Multiplication.gsp.** Tell students to work through steps 1–4 and do the Explore More tasks if they have time. Encourage them to ask a neighbor for help if they have questions about using Sketchpad.

7. Let pairs work at their own pace. As you circulate, listen to students' conversations. Here are some things to notice.

   • In worksheet step 2, students do not need to predict where the rabbit will land before entering the numbers and pressing *Jump Along.* The goal is for students to make a connection between the visual representation of the rabbit's jumps and the number sentences that describe how the rabbit jumped and where it landed.

   • To start a new problem in worksheet step 2, students should press *Erase Traces* and *Move Rabbit to 0.* For variety, students can change the color of the rabbit's traces by selecting the point below the rabbit and choosing **Display | Color** to pick a new color.

   • If students happen to enter numbers that cause the rabbit to jump beyond the visible portion of the screen, the extra jumps will not be visible if students scroll across after the rabbit has finished jumping.

   • In worksheet step 3, if any students are asking for help, prompt them to persist in doing their own thinking by asking, ***Have you experimented? Are you thinking about all the information given? What piece of information about the rabbit are you missing? Do you know how many jumps the rabbit takes? Do you know the size of each jump? Do you know where the rabbit lands?***

SUMMARIZE

Project the sketch. Expect to spend about 20 minutes.

8. Gather the class. Students should have their worksheets with them. Go to page "Jump Along." Discuss worksheet steps 3a and 3b. Use the model to display both five jumps of 2 and two jumps of 5 on the number line. Make the traces different colors.

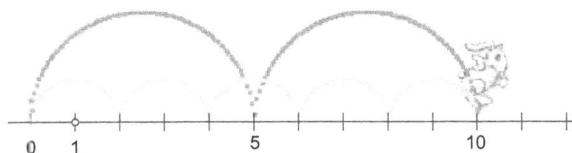

Number of Jumps = 2 | Jump Along |

Jump By = 5

What is the same and what is different about these trips the rabbit has taken? Take responses. Sample responses follow.

The numbers we put in are the same—2 and 5—but they're reversed.

The actions that the rabbit takes are related. In one problem, the rabbit takes five jumps of 2. In the other problem, the rabbit takes two jumps of 5.

The rabbit landed at the same place in both problems.

The number sentences look almost alike: $2 \times 5 = 10$ and $5 \times 2 = 10$.

It doesn't matter what order you multiply 2 and 5 in. The answer is the same.

Whenever you multiply two numbers, it doesn't matter what order you multiply them in. The product is always the same.

9. Write on the board the equation $6 \times 3 = ?$. ***Think of a story problem for the rabbit that goes with this number sentence.*** One possibility is this: The rabbit starts at 0 and takes six jumps of 3. Where does it get to?

   As another example, write the equation $4 \times ? = 20$. A story problem is this: The rabbit starts at 0. It wants to get to 20 in four jumps. It wonders how big each jump will need to be.

EXTEND

1. For students who are ready to work with larger numbers, you can rescale the number line by dragging the point sitting at 1 so that it is closer to 0.

2. Use the number line model to introduce students to the distributive property.

   • Set the rabbit to start at 0 and make 4 jumps of 3.

   • Record "4 jumps of 3" as well as 4×3 on the board.

   • With the rabbit at 12, take 2 more jumps of 3.

- Record "2 jumps of 3" as well as 2 × 3 on the board.

- Have students look at the trace of the rabbit's entire trip, and observe that altogether the rabbit took six jumps of 3, landing at 18. Elicit the idea that six jumps of 3 is another way of viewing four jumps of 3 plus two jumps of 3.

- Express this symbolically as 6 × 3 = 4 × 3 + 2 × 3.

3. ***What other questions can you ask about jumping along the number line?*** Encourage student curiosity. Here are some sample student queries.

How many different ways can the rabbit jump to 11, starting at 0?

What if our number sentence is 3 × 2 + 1 × 4 = 10? What story about the rabbit could go along with this number sentence?

What if the Jump By *number is less than 0? What happens then?*

ANSWERS

2.

The Rabbit Makes . . .	Number of Jumps	Jump By	Get To	Multiplication Number Sentence
a. 4 jumps of 2	4	2	8	4 × 2 = 8
b. 3 jumps of 4	3	4	12	3 × 4 = 12
c. 5 jumps of 2	5	2	10	5 × 2 = 10
d. 6 jumps of 3	6	3	18	6 × 3 = 18

3.

Problem	Number of Jumps	Jump By	Get To	Multiplication Number Sentence
a.	5	2	10	5 × 2 = 10
b.	2	5	10	2 × 5 = 10
c.	4	3	12	4 × 3 = 12
d.	3	4	12	3 × 4 = 12
e.	4	6	24	4 × 6 = 24
f.	6	4	24	6 × 4 = 24

4. Rows a and b have the same numbers, only "flipped." Row a shows that five jumps of size 2 land the rabbit at 10. Row b shows that two jumps of size 5 also land the rabbit at 10. The same relationship can be found between rows c and d and between rows e and f.

5. The rabbit can take five jumps to land on 20.

6. The rabbit can take four jumps to land on 28.

7. The rabbit can jump by 6 to land on 36.

8. The rabbit cannot reach the odd numbers.

9.

The Rabbit Makes . . .	Number of Jumps	Jump By	Get To	Number Sentence
a. 3 jumps of 2	3	2	8	$2 + 3 \times 2 = 8$
b. 5 jumps of 4	5	4	22	$2 + 5 \times 4 = 22$
c. 7 jumps of 3	7	3	23	$2 + 7 \times 3 = 23$

Jump Along

Name:

Help a rabbit jump along a number line.

EXPLORE

1. Open **Jump Along Multiplication.gsp.** Go to page "Jump Along."

2. Get the rabbit where it wants to go. Use the table to record.

Rabbit Starts at 0

The Rabbit Makes. . .	Number of Jumps	Jump By	Get To	Multiplication Number Sentence
a. 4 jumps of 2				
b. 3 jumps of 4				
c. 5 jumps of 2				
d. 6 jumps of 3				

3. Get the rabbit where it wants to go. Use the table on the next page to record.

   a. The rabbit wants to jump by 2's and get to 10.

   b. The rabbit wants to take 2 jumps and get to 10.

   c. The rabbit wants to jump by 3's and get to 12.

   d. The rabbit wants to take 3 jumps to get to 12.

   e. The rabbit wants to jump by 6's to get to 24.

   f. The rabbit wants to take 6 jumps to get to 24.

Rabbit Starts at 0

The Rabbit Makes. . .	Number of Jumps	Jump By	Get To	Multiplication Number Sentence
a.				
b.				
c.				
d.				
e.				
f.				

4. Look at your table in step 3. What patterns can you find?

_____

_____

_____

_____

EXPLORE MORE

5. How close can the rabbit get to 21 when it jumps by 4?

_____

6. How close can the rabbit get to 30 when it jumps by 7?

_____

7. How close can the rabbit get to 35 when it makes 6 jumps?

_____

8. If the rabbit jumps by 2's, what numbers is it not able to get to?

_____

Jump Along

continued

9. Go to page "Explore More."
   Drag the point below the rabbit so that it moves to 2.
   Show each problem on the number line. Record what happens.

Rabbit Starts at 2

The Rabbit Makes. . .	Number of Jumps	Jump By	Get To	Number Sentence
a. 3 jumps of 2				
b. 5 jumps of 4				
c. 7 jumps of 3				

New York City Title I Elementary School Activities with The Geometer's Sketchpad
© 2012 Key Curriculum Press

Sum and Product Puzzles:
Number Sense and Mental Computation

For GSP5

ACTIVITY NOTES

INTRODUCE

Project the sketch on a large-screen display for viewing by the class. Expect to spend about 10 minutes.

1. Open **Sum Product Puzzles Present.gsp** and go to page "How to Play 1." Say, ***In this activity, you'll use your knowledge of addition and multiplication facts to solve puzzles the computer gives you.*** Explain that this page and the next page tell students how to use the model. When they work on their own, students can refer to these pages to be reminded of what to do.

2. Go over the directions on this page and on page "How to Play 2." Explain these features of the puzzles.

   • The computer picks two numbers from 1 through 9. These two numbers may or may not be the same.

   • Students must determine the two numbers picked by the computer. The sum of the two numbers and the product of the two numbers are displayed on screen.

   • Students pick two numbers that they think solve the puzzle and drag them across the vertical divider line. If the sum of the two numbers matches the given sum, the corresponding rectangle turns green. Similarly, if the product of the two numbers matches the given product, the corresponding rectangle turns green. The problem is solved when both rectangles turn green.

   • If the two numbers students drag across the line do not give the sum and product, students drag one or both of them back to the left of the line and choose other numbers to try.

3. Go to page "Play the Game" and let students know this is the page they will use when they work independently.

   • Pressing *New Problem* moves numbers back to the left of the vertical line and creates the next randomly generated puzzle.

DEVELOP

4. Make sure students know where to locate **Sum Product Puzzles.gsp.**

5. As students work, you may want to observe how they approach the puzzles and the strategies they develop.

   • Some students may want to use paper and pencil. Let them use it once or twice, and then ask them to work mentally to solve the problems.

   • Are students fluent with the facts and comfortable solving the problems using mental computation?

   • Do they make separate lists of numbers that satisfy the given sum and the given product? If students are solving a puzzle in which the sum is 12 and the product is 27, their scratch work might look like this.

6 + 6 = 12	6 × 6 = 36	No
4 + 9 = 12	4 × 8 = 32	No
3 + 9 = 12	3 × 9 = 27	Yes !

   • Do they first look for two numbers that add to the given sum and then check what the product of those numbers is?

   • Do they first look for two numbers that multiply to the given product and then work with the factors to arrive at the given sum, a strategy that students with strong number sense are likely to employ?

SUMMARIZE

6. After some students have solved puzzles, provide an opportunity for them to talk together about the strategies they used.

EXTEND

1. Have student pairs create puzzles using page "Make Your Own."

2. Facilitate class discussion of these two questions.

   • *Can you name a sum and a product that would not work in this game?* [A sum of 4 and a product of 12, for instance, has no solution.]

   • *Do you think there will always be just one pair of numbers that works? Can there ever be more than one answer to a problem?* [By solving enough problems, students will sense (correctly) that there can be at most one pair of numbers that solve each problem.]

3. ***What other questions occurred to you about this kind of puzzle? You can ask a question you don't know the answer to.*** Encourage all student inquiry. Here are sample student queries.

   Could you solve the puzzles if there were three secret numbers instead of two?

   Is there a pattern to the number of pairs that have a given sum?

   Is there a pattern to the number of pairs that have a given product? For example, if the product is 4, there are two pairs of numbers that might solve the puzzle: 1, 4 and 2, 2.

   What product has the most possible pairs of the numbers in the game?

   What is the largest product possible for a pair of numbers in the game?

   Would all pairs of products and sums have solutions if you allowed fractions?

Comparing Fractions:
Number Sense and Benchmarks

ACTIVITY NOTES

INTRODUCE

Project the sketch for viewing by the class. Expect to spend about 10 minutes.

1. Distribute the worksheet and ask students to work on comparing the fractions in step 1 for a few minutes. (For now, they should ignore the write-on line beside each pair of fractions.) Explain that this is a warm-up for the exploration the class will do together. It's not expected that students will be able to compare all pairs.

2. Open **Comparing Fractions Number Sense.gsp** and go to page "Compare." To begin, make sure the model looks like the illustration at right. If students have not used a fraction-circle model before, ask questions to check that students are able to explain the relationships between the numerator of a fraction, its denominator, the number of parts in the circle, and the parts shaded. You might ask, **What would the top circle look like if I changed 1/4 to 1/8?** Take responses and then change the denominator of the top fraction by selecting the denominator with the **Arrow** tool and pressing **+** on the keyboard four times. As you do, let students know how you are changing the parameter for the denominator, and pause between each press.

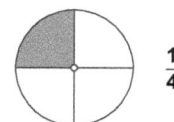

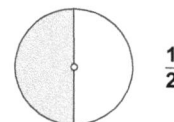

$\frac{1}{4}$

$\frac{1}{2}$

DEVELOP

Continue to project the sketch. Expect to spend about 35 minutes.

3. Facilitate discussion of the fraction pairs in steps 1a–g. Here are suggestions for using the model. Because visual models are being presented, make sure students know when the *numerical* representation is being discussed. Students are asked to write pairs of fractions on the worksheet. This provides an opportunity to check that students are connecting the visual examples with their fraction number sense and with the numerical representation of a fraction.

A. 2/9 AND 5/9

Using the model, show 2/9 in both circles and then increase the numerator to 5 in one fraction by selecting the numerator and pressing **+** three times. Pause between each press so that students can see the effect on the circle. **What can you say about comparing fractions that have the same denominator?**

Comparing Fractions: Number Sense and Benchmarks
continued

You may wish to introduce the term *common denominators* now and use it in this activity.

When the denominators are the same, it's easy to compare the fractions. The parts are the same size, so the fraction representing more parts—the fraction with the larger numerator—is greater. With the numerator 5 selected, press **+** until the numerator is 8 or 9 to further illustrate this.

In worksheet step 1a, have students write another pair of fractions that have the same denominator and different numerators. Then have students use the symbols for greater than and less than to show which fraction is greater.

B. 4/7 AND 4/11

Using the model, show 4/7 in both circles and then, in one fraction, increase the denominator to 11 by selecting the denominator and pressing **+** four times. Pause between each press so that students can see the effect on the circle. *What can you say about comparing fractions that have the same numerator?*

When the numerators are the same, it is also easy to compare the fractions. Each fraction has the same number of equal parts, but the size of the parts differs. The more parts, the smaller the parts are. The fraction with the larger denominator is less than the fraction with the smaller denominator.

Exploring extreme cases is worthwhile if the result is an engaging and memorable demonstration of an important concept.

Students may propose using the model to increase the denominator to even larger numbers. Select the larger denominator and, pressing the **+** key repeatedly, go all the way to 50.

On the worksheet, students should write another pair of fractions that have the same numerator and different denominators, and indicate which fraction is greater.

C. 2/4 AND 5/10

A denominator of 100 makes it impossible to see the divisions (which students enjoy seeing), whereas a denominator of 50 allows students to discern the divisions.

Students should recognize that both of these familiar fractions are equivalent to 1/2. Promote discussion about fractions that are equivalent to 1/2. Test a number of cases using the model, allowing time for students to conclude that for fractions equivalent to 1/2, the denominator is twice the numerator. Students may again suggest trying some very large values for the denominator. The quick way to change a parameter is to double-click the number, enter a new value in the dialog box that appears, and click OK. Use this method when you aren't interested in showing incremental changes.

Students should write another pair of fractions that are equivalent to 1/2, and, therefore, to each other, and use the equal sign to show the relationship.

D.   9/16 AND 4/9

Neither the numerators nor the denominators are the same. Another strategy is needed. A sample student response is this one: *I know that 4/9 is less than 1/2 because 4 is less than half of 9, and I know that 9/16 is larger than 1/2 because 9 is more than half of 16. So, 9/16 is greater.* If no student suggests a successful strategy, start the class off by asking whether 4/9 is greater than or less than 1/2.

When ideas have been considered, model 4/9 in one fraction circle by changing the numerator to 1 and the denominator to 9. Then select the numerator and press **+** until the numerator reaches 4. Stop and ask for students' thinking. ***Are you convinced that 4/9 is less than 1/2? What will the model look like if I press + again?*** Press **+** again to show 5/9. Then press **−** to return to 4/9.

Model 9/16 in the other fraction circle, changing the numerator to 1 and the denominator to 16 to start. Select the numerator and press **+** until the numerator is 9. Again discuss the results.

Students should write another pair of fractions for which relating the fractions to 1/2 is a handy way to compare them. Invite a few students to share the fraction pairs they have written, and model the fractions. (Some students may write pairs of fractions for which thinking about equivalence is an equally handy strategy, for example, 2/5 and 7/10. If students propose this alternate strategy, take time to discuss it.)

E.   2/21 AND 7/8

Thinking about how close the fractions are to benchmarks is again useful. This time students may compare the given fractions to the benchmarks 0 and 1, explaining that 2/21 is close to 0, and 7/8 is close to 1. Ask students to explain how they know that 2/21 is close to 0, and how they know 7/8 is close to 1. ***What can you say about fractions that are close to zero?*** [The numerator is much smaller than the denominator.] ***What can you say about fractions that are close to 1?*** [The numerator is close to the denominator.]

In this model, increasing the numerator until the fraction is greater than 1 will not result in the corresponding picture. Go to page "Improper Fractions" to work with fractions greater than 1.

Show 21/21 in one fraction circle, select the numerator and press −
repeatedly until the circle shows 2/21. In the other circle, set the fraction
to 1/8 and increase the numerator repeatedly until it shows 7/8.

Students should write another pair of fractions for which thinking
about the benchmarks 0 and 1 is useful. Invite students to share their
fraction pairs.

F. 3/4 AND 5/6

Some students may propose that the fractions are equivalent because
each is "*missing one part.*" Show the fractions in the model by double-
clicking each parameter and changing its value in the dialog box.
Students will see that the fractions are not equivalent. ***Does that make
sense? Why aren't the fractions equal to each other? After all, each
whole is missing exactly one part.***

Direct students' attention to the model and ask them to observe just one
of the fraction circles. Select *both* the numerator and the denominator
and press **+** on the keyboard, causing the numerator and denominator
to increase by one simultaneously. Repeat several times, pausing after
each press. ***What can you say?*** Note whether students are convinced that
as the number of parts increases, the size of the missing part decreases.

Elicit the idea that when the missing part is bigger, the given fraction is
smaller. Students should reason that 5/6 is greater than 3/4 because 5/6
is 1/6 less than 1, whereas 3/4 is 1/4 less than 1.

Students should write another pair of fractions for which thinking
about the "missing part" is useful.

G.   8/8 AND 9/13

Students can compare again using the benchmark 1. ***What can
you say about fractions equal to 1?*** [The numerator is equal to the
denominator.] Students should recognize that 8/8 is equal to 1 and 9/13
is less than 1.

Students may want to model, using the sketch, other pairs in which one
fraction is equal to 1 and the other is less than 1.

4. To include discussion of fractions greater than 1, go to page "Improper
   Fractions." Make sure the models show 1/6 and 1/8 to start. Write the
   fractions 9/6 and 16/8 on the board, and ask students to think about
   how they compare. Provide thinking time.

For GSP5 ACTIVITY NOTES

What do you think the model will look like if we show 9/6 and 16/8? Select the numerator in 1/6 and press **+** until 9/6 is shown. Select the numerator in 1/8 and press **+** until 16/8 is shown. *How can you tell that a fraction is greater than 1?* [The numerator is greater than the denominator.] *How can you tell that a fraction is between 1 and 2?* [The numerator is greater than the denominator, but less than twice the denominator.]

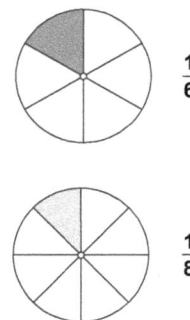

$\frac{1}{6}$

$\frac{1}{8}$

If time allows, have students suggest pairs of fractions to compare in which at least one of the fractions is greater than 1. Discuss students' thinking about the comparisons, and model the fractions.

SUMMARIZE

Working with or without the projected sketch, expect to spend about 15 minutes.

5. Point out that students have devised a number of strategies for comparing fractions. Focus the discussion on the usefulness of looking for a handy, or efficient, strategy when comparing fractions. Ask students to take a few minutes to consider the fractions in worksheet step 2. *Think about which strategy helps you compare the two fractions in each pair most easily and quickly.*

When most students have had a chance to consider all the pairs, ask about each pair, inviting students to share the strategies they found easiest to use.

(Note that, in general, students should not think that they need to agree on a most efficient strategy when thinking about numbers and computation; what is most efficient depends upon an individual's number sense and experience with computation.)

EXTEND

1. Have students work independently in pairs using pages "Compare" and "Improper Fractions." Give these directions:

   • One student chooses a pair of fractions to compare, trying to stump the other. The student must know which fraction is greater.

   • The second student chooses the fraction he or she thinks is greater.

   • The partners explain their reasoning and model the fractions to check.

   • Students can keep score on paper.

New York City Title I Elementary School Activities with The Geometer's Sketchpad
© 2012 Key Curriculum Press

2. Page "Match Up" provides another model for independent practice comparing fractions. The sketch randomly generates pairs of fractions and shows them numerically and in fraction-circle models. Students should determine which fraction is greatest by using the numerical representation as well as by deciding which fraction-circle model shows which fraction. For a hint, students can press *Show Lines*. They can also compare the fraction circles directly by selecting the center of one circle and dragging it onto the center of the other circle.

3. Present three or more fractions and ask students to put them in order. Some students may find it helpful to use the model.

Compare the Fractions

For GSP5 Name: _____

Find handy ways to compare fractions.

EXPLORE

1. Compare the fractions in each pair. Use the symbols >, <, or = to show how they compare.

   a. 2/9   5/9   _____

   b. 4/7   4/11   _____

   c. 2/4   5/10   _____

   d. 9/16   4/9   _____

   e. 2/21   7/8   _____

   f. 3/4   5/6   _____

   g. 8/8   9/13   _____

2. For each pair of fractions, think about which strategy helps you compare the fractions quickly and easily.

   a. 6/11   24/50

   b. 8/9   5/6

   c. 3/12   8/12

   d. 30/60   7/14

   e. 40/50   4/4

   f. 7/17   7/100

   g. 11/12   2/15

Fraction Tiles: Equivalent Fractions

INTRODUCE

Project the sketch for viewing by the class. Expect to spend about 10 minutes.

1. Present the following problem. *Joe and Jesse used the same-sized square baking pans to make two pans of brownies, one for Joe and one for Jesse. Joe cut his pan into four equal pieces and ate one. Jesse said, "I'm really hungry—one piece isn't going to be enough for me!" So he cut his pan in eight equal pieces and ate two. Who ate more?*

   This problem should be accessible to students. Elicit students' prior knowledge of equivalent fractions. Students should point out that the fractions $\frac{1}{4}$ and $\frac{2}{8}$ are two names for the same amount of brownies; they are *equivalent fractions.*

2. Open **Fraction Tiles Equivalent Fractions.gsp** and go to page "Example 1." Point out the equivalent fractions shown, $\frac{3}{5} = \frac{9}{15}$. *The action buttons here show what the grids and tiles in this model look like. I'll press them and we'll talk about what you see.*

   • Press *Show Fifths Grid.* Students will observe that the square represents one whole, and that the grid is divided into fifths, the part named by the denominator of the fraction $\frac{3}{5}$.

   • Press *Show $\frac{1}{5}$ Tiles.* Students will observe that the number of shaded parts represents the number of fifths in $\frac{3}{5}$, or the numerator of the fraction.

   • Press *Show Thirds Grid.* Give students time to recognize that when one grid is crossed by another, the type of part and the number of parts (the denominator and the numerator) change without changing the part of the whole that is shaded (the value of the fraction).

3. *You will use the custom tools to make your own grids and tiles as you model equivalent fractions.* Go to page "Example 2." *Let's use the Sketchpad fraction model to show that these fractions are equivalent.*

   • In the Toolbox, press and hold the **Custom** tool icon to display the Custom Tools menu. Point out the **Grid Horizontal** and **Grid Vertical** tools. *The grid tools will divide the square vertically or horizontally into equal parts. We'll use one of them to show the square divided into fourths.*

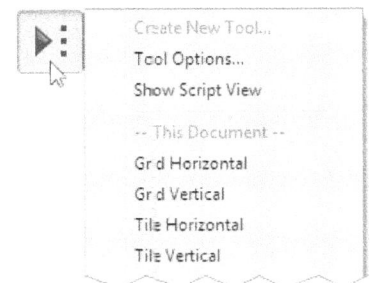

   Create New Tool...
   Tool Options...
   Show Script View

   -- This Document --
   Grid Horizontal
   Grid Vertical
   Tile Horizontal
   Tile Vertical

Fraction Tiles: Equivalent Fractions

continued

ACTIVITY NOTES

Choose **Edit | Undo** to back up if you mistakenly place the wrong grid.

(You can let students choose which grid to use. The directions below assume the vertical grid, but you can use the horizontal grid as easily.)

Choose the **Grid Vertical** tool. *We have to click a denominator to tell the tool how many parts to make.* In the sketch, click the denominator 4 in the fraction $\frac{1}{4}$. *We click the top-left corner of the first square to tell the tool where to put the grid. Move your pointer over the top-left corner of the square. When the corner point is highlighted, click the point.* The grid appears.

- *Let's show $\frac{1}{4}$.* Display the Custom Tools menu again and point out the **Tile Horizontal** and **Tile Vertical** tools. *The tile tools shade parts of grids, either vertically or horizontally. Which kind of tile should I choose?* [Vertical]

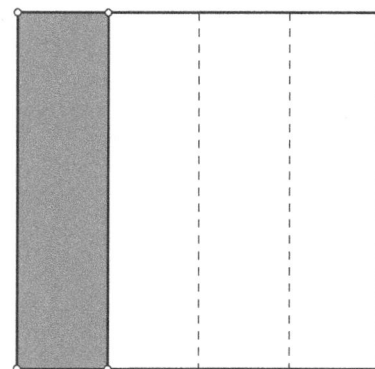

  Choose the **Tile Vertical** tool. *We also have to click a denominator for this tool.* In the sketch, click the denominator 4 in the fraction $\frac{1}{4}$. Then click the point at the top-left corner of the square to place the tile.

- *Can we show eighths? How?* Some students will suggest dividing the square in half horizontally. Choose the **Grid Horizontal** tool. Place a grid by clicking the denominator 2 of the fraction $\frac{1}{2}$ in the sketch and then clicking the top-left corner of the square.

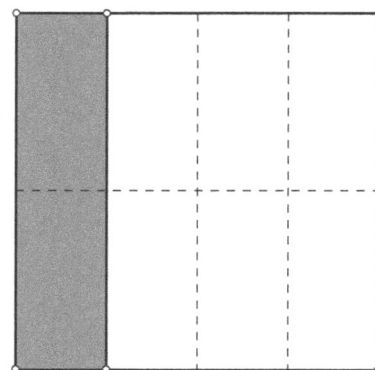

4. Go to page "a b" of the sketch. *Starting on this page, you'll see two separate exercises per page. You'll decide which of the grid and tile tools to use to represent the equivalent fractions.*

   Go to page "c d." *On some pages, you'll need to find a missing numerator or denominator in order to make the fractions equivalent.*

174

New York City Title I Elementary School Activities with The Geometer's Sketchpad
© 2012 Key Curriculum Press

5. Students should record their answers on the worksheet and/or by editing the caption above the square for each exercise. Model using the **Text** tool to edit a caption if this skill is new to students.

6. If you want students to save their Sketchpad work, demonstrate choosing **File | Save As,** and let them know where to save.

DEVELOP

Expect students working at computers to spend 40 minutes.

7. Assign students to computers and tell them where to locate **Fraction Tiles Equivalent Fractions.gsp.** Distribute the worksheet. Tell students to do steps 1 through 3, and the Explore More if they have time. Encourage students to ask a neighbor for help if they have questions about using Sketchpad.

8. Let students work at their own pace. Here are some things to be aware of as you observe and ask students to share their thinking.

   • Encourage students to review pages "Example 1" and "Example 2" on their own before going on to page "a b."

   • Note students' strategies for choosing the second grid when a numerator or denominator is missing. Students may be able to find the missing values without using the model. It's important that they also are able to reason about what grids to use to model the equivalence.

   • When you observe students having difficulty choosing the right grid for an exercise, ask guiding questions such as the following in order to encourage them to reason on their own.

      You have shown thirds with vertical divisions. With your finger, show me how you would "cut the cake" so that there are sixths. Okay. Which grid do you think would do that?

      You're wondering how to find the missing denominator in this problem. What do you know? Instead of 2 thirds, there will be 8 of some other part. Instead of 2 shaded pieces, there will be 8. There will be four times as many shaded pieces. What can you do with that information?

9. Students who have time should do the Explore More.

10. If students will save their Sketchpad work, remind them to do so now.

SUMMARIZE

Project the sketch. Expect to spend about 25 minutes.

11. Bring the class together for discussion. Invite students to the computer to model problems they found most interesting or challenging.

12. To extend students' thinking and assess their understanding of equivalent fractions, ask, ***What are some denominators you can't make with these tools? Why can't you make them?*** Facilitate a discussion, encouraging students to suggest and evaluate denominators. A student might propose, *You can't make 11ths. There are no 11ths grids or tiles, and no combination of a vertical and a horizontal grid will make 11ths.* If students need prompting for exploring why some denominators cannot be made with the model, ask questions such as the following.

    Can anyone explain why no two grids can be used to make 11ths?

    Are there other denominators that are not possible for the same reason?

    It's not as important that students know all the reasons some denominators are not possible as it is that they persevere in doing their own thinking about this rich question. You might offer this question as a writing prompt and give students time to think more about the question, in pairs or groups, and then respond individually.

    For your information, here are the denominators that can't be made with this model.

    - Denominators that are prime numbers greater than 7. Grids and tiles for primes greater than 7 are not provided in the custom tools. A prime number has no factors other than 1 and itself, so no combination of a vertical grid and horizontal grid (one number of parts multiplied by another number of parts) can produce a part that is a prime number.

    - Denominators larger than 100. Using the 10ths vertical grid and the 10ths horizontal grid together divides the whole into 100ths. No grids are provided that divide the square into more than 100 parts.

    - Denominators whose factor pairs are numbers that are not both provided in the grids. Two examples are 77 and 51. The factors pairs for 77 are 1 and 77, 7 and 11; the model provides a 7ths grid, but not an 11ths grid. The factor pairs for 51 are 1 and 51, 3 and 17; the model provides a 3rds grid, but not a 17ths grid.

New York City Title I Elementary School Activities with The Geometer's Sketchpad
© 2012 Key Curriculum Press

13. If students had time to work with the Explore More tasks and time allows, have them share some of the problems they created and talk about solving others' problems.

EXTEND

1. *What questions about equivalent fractions occur to you?* Encourage curiosity. Questions of mathematical interest include these.

   • How many different ways can you shade (eat) two pieces of the eight in Jesse's cake? How many ways if the two pieces must be adjacent?

   • Could you fill in an equation for equivalent fractions if two spaces were blank, for example the numerator of one fraction and the denominator of the other?

   • Are there fractions to which no other fraction is equivalent?

   • Can you tell whether fractions are equivalent without shading areas?

2. The challenges on page "Extend 1" preview the concept of fractions expressed in lowest terms. Students don't need to learn this term now. Provide these directions.

   Using one tile, shade in 4 of the 12 parts in the square on the left. Tell what tile you used. Is there more than one way?

   Use one tile to shade in 4 of the 12 parts in the square on the right. Tell what tile you used. Is there more than one way?

   Note any reasoning and strategies you see students using. If some students need help getting started, choose one model to explore with them and ask questions such as these.

   How could you shade $\frac{4}{12}$ using a tile more than once?

   What are all the ways you could shade $\frac{4}{12}$ with tiles in this model?

   Provide time for students to discuss their solutions, using their own language as they describe finding the "one-tile solution" (the tile that represents a given fraction in the lowest terms).

   On this page students must use a vertical thirds tile in the square on the left, and a horizontal thirds tile in the square on the right.

3. On page "Extend 2," students can make up problems like the one on page "Extend 1" and have other students solve them. Reinforce that it must be possible to shade the fraction using only one tile.

ANSWERS

3. See examples of possible solutions in **Fraction Tiles Equivalent Fractions Present.gsp.**

   c. $\dfrac{2}{3} = \dfrac{6}{9}$ d. $\dfrac{1}{4} = \dfrac{4}{16}$ e. $\dfrac{2}{3} = \dfrac{8}{12}$

   f. $\dfrac{1}{4} = \dfrac{3}{12}$ g. $\dfrac{1}{3} = \dfrac{4}{12}$ h. $\dfrac{2}{5} = \dfrac{6}{15}$

   i. $\dfrac{2}{3} = \dfrac{10}{15}$ j. $\dfrac{1}{5} = \dfrac{3}{15}$ k. $\dfrac{3}{10} = \dfrac{6}{20}$

   l. $\dfrac{5}{6} = \dfrac{15}{18}$

4. Problems will vary.

New York City Title I Elementary School Activities with The Geometer's Sketchpad
© 2012 Key Curriculum Press

Equivalent Fractions

For GSP5

Name:

Use an area model to find equivalent fractions.

EXPLORE

1. Open **Fraction Tiles Equivalent Fractions.gsp.**
   To choose a custom tool, press and hold the **Custom** tool icon, and then choose the tool from the menu.

2. On page "a b," show that the fractions are equivalent.

3. On pages "c d" through "k l," build the models. Record the missing numbers here.

   c. $\dfrac{2}{3} = \dfrac{}{9}$ d. $\dfrac{1}{4} = \dfrac{}{16}$

   e. $\dfrac{2}{3} = \dfrac{8}{}$ f. $\dfrac{1}{4} = \dfrac{}{12}$

   g. $\dfrac{1}{3} = \dfrac{4}{}$ h. $\dfrac{}{5} = \dfrac{6}{15}$

   i. $\dfrac{}{3} = \dfrac{10}{15}$ j. $\dfrac{1}{5} = \dfrac{}{15}$

   k. $\dfrac{3}{} = \dfrac{6}{20}$ l. $\dfrac{5}{} = \dfrac{15}{18}$

EXPLORE MORE

4. Go to page "Make Your Own." Make up two problems for others to solve. One problem should have a missing numerator. The other should have a missing denominator.

   To type your fractions, double-click in the equation above each square. Delete the question marks, and type your numbers.

   Record your problems and their solutions here.

   m.   n.

Jump Along: Equivalent Fractions on the Number Line

INTRODUCE

Project the sketch for viewing by the class. Expect to spend about 20 minutes.

Notice that the top number line represents 1 as $\frac{2}{2}$.

Although it is possible to enter a non-integer value for the *Number of Jumps* parameter, students will use only integer values in this activity.

1. Open **Jump Along Equivalent Fractions.gsp.** Go to page "Jump to 1/2." Explain, *Today you'll be using number lines to explore fractions. You'll direct the rabbit to jump along the number line. There are two ways to control the rabbit: You can change how many jumps it will make, and you can change the size of each jump.*

2. Enlarge the window so that it fills most of the screen. Students will be focusing on the interval from 0 to $\frac{2}{2}$. To make this interval easier to see, drag the point at $\frac{2}{2}$ to the right.

3. Ask students to predict what will happen when you press the *Jump Along* button. Take responses and then press the button. The rabbit takes 1 jump (because *Number of Jumps* is equal to 1) and the size of the jump is $\frac{1}{2}$ (because *Jump By* is equal to $\frac{1}{2}$). The rabbit lands at $\frac{1}{2}$, leaving a trace of its jump.

4. Distribute a copy of the Jump Along table from the worksheet to each student. Explain that students will be finding different ways for the rabbit to reach $\frac{1}{2}$. They should fill in the first row of the table with a "1" in the Number of Jumps column because one jump of $\frac{1}{2}$ landed the rabbit at $\frac{1}{2}$.

5. Using the **Arrow** tool, select the denominator of the *Jump By* fraction. Press the **+** key on your keyboard once to increase the value of the denominator by one. The value of *Jump By* is now $\frac{1}{3}$. Ask students what happened to the fractions along the top number line. Students should note that the top number line is now divided into thirds. The bottom number line has not changed. For students who need to see the prior number line again, press the **−** key on your keyboard to decrease the denominator by 1.

6. With *Jump By* set at $\frac{1}{3}$, ask students what number of jumps will land the rabbit at $\frac{1}{2}$. By looking at the two number lines, some students may reason that one jump of $\frac{1}{3}$ is not enough, while two jumps of $\frac{1}{3}$ is too much. Other students may say that the number of jumps must be half of 3, but 1.5 is not a whole number and only whole numbers are allowed. Because it is not possible to reach $\frac{1}{2}$ with jumps of size $\frac{1}{3}$, tell students to place an **X** in the Number of Jumps column of the table.

The rabbit always starts its jumps at 0. Pressing *Reset* moves the rabbit back to 0 without erasing the traces.

7. Change *Jump By* to $\frac{1}{4}$, discuss what happened to the markings along the top number line, and ask the class how many jumps are needed to land the rabbit at $\frac{1}{2}$. Before testing to see that two jumps of $\frac{1}{4}$ works, select the point below the rabbit and choose **Display | Color** to pick a different color. Now, when you press *Jump Along*, the trace of the rabbit's jumps will be in the new color.

8. Explain that students will be trying the other *Jump By* numbers in the table to see which ones allow the rabbit to land at $\frac{1}{2}$. ***Be on the lookout for patterns! Investigate the traces on screen and the numbers in the table.***

DEVELOP

Expect students at computers to spend about 45 minutes.

9. Assign students to computers and tell them where to locate **Jump Along Equivalent Fractions.gsp.** Distribute the worksheet, including multiple copies of the table. Tell students to work through steps 1–9, using a new copy of the table for each new landing spot. Encourage students to ask a neighbor for help using Sketchpad if they have questions.

10. Let pairs work at their own pace. As you circulate, listen to students' conversations. Here are some things to notice.

   • In worksheet step 2, students may use the two number lines together to tell whether it is possible to reach $\frac{1}{2}$ for a particular *Jump By* number. When *Jump By* equals $\frac{1}{5}$, for example, none of the one-fifth intervals on the top number line coincide with $\frac{1}{2}$ on the bottom line. Thus no number of one-fifth jumps will land the rabbit at $\frac{1}{2}$. Other students may figure out how to land at $\frac{1}{2}$ by finding equivalent fractions such as $\frac{2}{4}, \frac{3}{6}, \frac{4}{8}$, and so on. Knowing that $\frac{1}{2}$ is the same as $\frac{3}{6}$, for example, helps students to see that three jumps of $\frac{1}{6}$ will work.

Jump Along: Equivalent Fractions on the Number Line
continued

It is not possible to erase one trace without erasing all the traces.

- In worksheet step 2, the rabbit traces out ways to get to $\frac{1}{2}$. Ask students what they notice. The number of jumps needed to reach $\frac{1}{2}$ increases by one for each trip in the table. When the rabbit jumps twice to reach $\frac{1}{2}$, it divides the interval from 0 to $\frac{1}{2}$ into two equal parts. When the rabbit jumps three times to reach $\frac{1}{2}$, it divides the same interval into three equal parts, and so on.

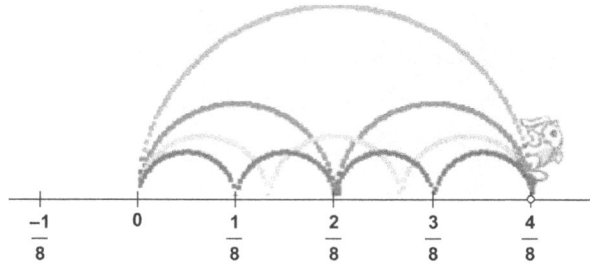

This is a good place to stop on the first day.

- In worksheet step 3, ask students what patterns they notice in their tables. Students may say that they can't make the rabbit jump to $\frac{1}{2}$ when the denominator of the *Jump By* fraction is an odd number. They may also say that the *Number of Jumps* value keeps increasing by one as they look down the column in the table.

There are many ways to represent $\frac{1}{4}$ as a fraction. Equivalent fractions such as $\frac{2}{8}$ and $\frac{3}{12}$ will also work.

11. Introduce worksheet steps 10–12. Now students will keep the number of jumps the same (2 jumps) and enter different *Jump By* values that land the rabbit at $\frac{1}{2}$, at 1, and at $\frac{4}{3}$.

12. As you circulate, listen to students' conversations. Some students might say, *We think there's only one way to reach $\frac{1}{2}$. Take 2 jumps of $\frac{1}{4}$.* These students are correct in thinking that only a jump whose size is $\frac{1}{4}$ will work. Extend their thinking by asking, **Is there another way to tell the rabbit to take a jump of $\frac{1}{4}$?** Let students think about this question and talk with neighbors if they need to.

SUMMARIZE

Project the sketch. Expect to spend about 25 minutes.

13. Gather the class. Students should have their worksheets with them. Direct students' attention to the table they completed in worksheet step 2. **You found that 2 jumps of $\frac{1}{4}$ landed the rabbit at $\frac{1}{2}$. Can we write this information as a number sentence?** Work with students to develop the number sentences $\frac{1}{4} + \frac{1}{4} = \frac{1}{2}$ and $\frac{1}{2} = \frac{1}{4} + \frac{1}{4}$. Some students may also suggest $2 \times \frac{1}{4} = \frac{1}{2}$.

14. ***The name of this activity is Jump Along: Equivalent Fractions on the Number Line. Why do you think the title mentions equivalent fractions?*** Take responses. Students may say that there are lots of ways to reach a fraction like $\frac{1}{2}$; other fractions are also names for the same location on the number line. Some of these are $\frac{2}{4}, \frac{3}{6}, \frac{4}{8}, \frac{5}{10}$, and $\frac{6}{12}$. As a different example, each table in worksheet steps 10–12 contains a list of equivalent fractions.

EXTEND

What other questions can you ask about jumping along the number line? Encourage student curiosity. Here are some sample student queries.

In how many ways can we reach $\frac{6}{8}$ if we make the denominator of Jump By *equal to 8?*

In how many ways can we reach $\frac{12}{15}$ if we make the denominator of Jump By *equal to 15?*

Are there locations along the number line we can't reach using the Jump By *fractions in the* Jump Along *table?*

How close can we get to $\frac{1}{2}$ without exactly reaching it? How close can we get to 1 without exactly reaching it?

What happens if we make the Jump By *number negative?*

ANSWERS

2. 1 jump of $\frac{1}{2}$, 2 jumps of $\frac{1}{4}$, 3 jumps of $\frac{1}{6}$, 4 jumps of $\frac{1}{8}$, 5 jumps of $\frac{1}{10}$, and 6 jumps of $\frac{1}{12}$

3. The number of jumps needed to land the rabbit at $\frac{1}{2}$ increases by 1, from 1 to 6. Only *Jump By* fractions with even denominators allow the rabbit to land at $\frac{1}{2}$. The number of jumps is always half the denominator of the *Jump By* fraction.

4. 3 jumps of $\frac{1}{4}$, 6 jumps of $\frac{1}{8}$, and 9 jumps of $\frac{1}{12}$

5. The number of jumps needed to land the rabbit at $\frac{3}{4}$ is always a multiple of 3. The denominators of the *Jump By* fractions are all multiples of 4.

6. 2 jumps of $\frac{1}{2}$, 3 jumps of $\frac{1}{3}$, 4 jumps of $\frac{1}{4}$, 5 jumps of $\frac{1}{5}$, and so on

7. It is possible to land at 1 for any of the *Jump By* fractions. The number of jumps is always equal to the denominator of the *Jump By* fraction.

8. 4 jumps of $\frac{1}{3}$, 8 jumps of $\frac{1}{6}$ and 12 jumps of $\frac{1}{9}$

9. The number of jumps needed to land the rabbit at $\frac{4}{3}$ is always a multiple of 4. The denominators of the *Jump By* fractions are all multiples of 3.

10. *Jump By* values of $\frac{1}{4}$, $\frac{2}{8}$, $\frac{3}{12}$, and $\frac{4}{16}$ all land the rabbit at $\frac{1}{2}$. All of these values are equivalent to $\frac{1}{4}$.

11. *Jump By* values of $\frac{1}{2}$, $\frac{2}{4}$, $\frac{6}{9}$, and $\frac{4}{16}$ all land the rabbit at 1. All of these values are equivalent to $\frac{1}{2}$.

12. *Jump By* values of $\frac{2}{3}$, $\frac{4}{6}$, $\frac{6}{9}$, and $\frac{8}{12}$ all land the rabbit at $\frac{4}{3}$. All of these values are equivalent to $\frac{2}{3}$.

14. Problems will vary.

Jump Along Fractions

Name:

Find ways the rabbit can jump to a number on the number line.

1. Open **Jump Along Equivalent Fractions.gsp**. Go to page "Jump to 1/2."

   A rabbit is set to jump along the number line.

   To change the size of the jumps and the number of jumps, select a number and press + or − on your keyboard.

2. Use a Jump Along table. For each *Jump By* number, try to land the rabbit on $\frac{1}{2}$.

   If the rabbit lands on $\frac{1}{2}$, record the number of jumps.

   If the rabbit can't land on $\frac{1}{2}$, put an X.

   Look for patterns that will help you get the rabbit to $\frac{1}{2}$.

   Make each set of jumps a different color. Select the point below the rabbit and choose a new color using **Display | Color.**

3. Look at your table and the jumps traced by the rabbit. What patterns do you see?

   _____

   _____

   _____

4. Go to page "Jump to 3/4." Now, the rabbit must land on $\frac{3}{4}$. Use a new table. Try all of the *Jump By* numbers. Record your findings.

5. Look at your table and the jumps traced by the rabbit. What patterns do you see?

   _____

   _____

   _____

For
GSP5

6. Go to page "Jump to 1." Now the rabbit must reach 1.
   Record your findings in a new table.

7. Look at your table and the jumps traced by the rabbit.
   What patterns do you see?

   _____

   _____

   _____

8. Go to page "Jump to 4/3." Now, the rabbit must reach $\frac{4}{3}$.
   Record your findings in a new table.

9. Look at your table and the jumps traced by the rabbit.
   What patterns do you see?

   _____

   _____

   _____

In steps 10, 11, and 12, the rabbit must take exactly two jumps.

10. Go back to page "Jump to 1/2." Find four ways the rabbit can land
    on $\frac{1}{2}$. Record them in this table.

    Land on 1/2

Jump By	Number of Jumps
	2
	2
	2
	2

New York City Title I Elementary School Activities with The Geometer's Sketchpad
© 2012 Key Curriculum Press

For GSP5

11. Go to page "Jump to 1." Find four ways the rabbit can land on 1.

Land on 1

Jump By	Number of Jumps
	2
	2
	2
	2

12. Go to page "Jump to 4/3." Find four ways to land on $\frac{4}{3}$.

Land on 4/3

Jump By	Number of Jumps
	2
	2
	2
	2

EXPLORE MORE

13. Go to page "Explore More." Make a jump-along problem.

    Choose a place on the number line to land on and name it as a fraction. Decide how many jumps and the jump size.

    For practice, pretend you'd like to jump to $\frac{3}{5}$. Change the *Number of Divisions* to 5. Now the bottom number line will be divided into five equal parts between 0 and 1.

    A Make a caption "3/5" and drag it below the point at $\frac{3}{5}$. Use this point to help you figure out ways the rabbit can jump to $\frac{3}{5}$.

14. On the back of this sheet, record your jump-along problem and ways you solved it.

JUMP ALONG

Land on _____

Jump By	Number of Jumps
$\frac{1}{2}$	
$\frac{1}{3}$	
$\frac{1}{4}$	
$\frac{1}{5}$	
$\frac{1}{6}$	
$\frac{1}{7}$	
$\frac{1}{8}$	
$\frac{1}{9}$	
$\frac{1}{10}$	
$\frac{1}{11}$	
$\frac{1}{12}$	

Land on _____

Jump By	Number of Jumps
$\frac{1}{2}$	
$\frac{1}{3}$	
$\frac{1}{4}$	
$\frac{1}{5}$	
$\frac{1}{6}$	
$\frac{1}{7}$	
$\frac{1}{8}$	
$\frac{1}{9}$	
$\frac{1}{10}$	
$\frac{1}{11}$	
$\frac{1}{12}$	

New York City Title I Elementary School Activities with The Geometer's Sketchpad

Angle Measurement: Estimation Practice

INTRODUCE

Project the sketch on a large-screen display for viewing by the class. Expect to spend about 10 minutes.

1. Open **Angle Estimation.gsp.** Go to page "The Model." *In this activity you'll use a Sketchpad model to sharpen your skills at estimating the size of angles.* Explain that this page introduces the model. Invite a volunteer to follow the directions on the page as the class observes.

2. Go to page "How to Play 1." Explain that this page and the next page tell students how to use the model. When they work on their own, students can refer to these pages to be reminded of what to do.

   Invite a volunteer to follow the directions on this page and on page "How to Play 2" as the class observes.

3. Distribute the worksheet and explain how students will use it to record the work they do independently at computers.

DEVELOP

Expect students at computers to spend 15 to 30 minutes.

4. Make sure students know where to locate **Angle Estimation.gsp** when they work independently.

5. As students work, you may want to observe how they approach the estimation challenges.

   - Do they think about benchmark angles?

   - Do they tend to orient the angles with one side horizontal or vertical, or are they comfortable with different orientations?

   - Do they estimate reflex angles by estimating the size of the smaller angle and subtracting that number from 360°?

   - Are they refining and improving their estimates as they receive feedback from the computer?

SUMMARIZE

Project the sketch on a large-screen display for viewing by a small group. Expect to spend 10 to 15 minutes.

6. After some students have used the sketch, provide an opportunity for them to discuss the strategies they used and to assess whether they improved in their ability to estimate angle measurements.

How Close Are You?

For GSP5 Name:

Sharpen your skills at estimating angle sizes.

1. Open **Angle Estimation.gsp.**

2. Go to page "Estimate 1." For each angle you make, record your estimate in the first table. Then record the actual measurement.

3. Go to page "Estimate 2." In the second table, record the measurement you are trying to show. Then record the actual measurement of the angle you make.

Your Estimate	Actual Measurement	Measurement to Estimate	Actual Measurement

New York City Title I Elementary School Activities with The Geometer's Sketchpad
© 2012 Key Curriculum Press

Angle Measurement: Introducing Protractors

INTRODUCE

Project the sketch for viewing by the class. Expect to spend about 20 minutes.

1. Open **Introducing Protractors.gsp.** Go to page "180 A." Enlarge the document window so it fills most of the screen.

2. Explain, *Today you are going to measure angles using a tool called a protractor. Let's review what you know about angles.* Refer to the angle shown in the sketch. *What names could we give to this angle?* [Angle *ABC*, angle *CBA*, or angle *B*] Remind students that an angle is named by its vertex or by a point on one side of the angle, the vertex, and a point on the other side of the angle. Emphasize that the vertex must always be the second point when three points are named.

More precise angle measures are given in degrees, minutes, and seconds: 30° 35′ 10″.

3. Point out the arc symbol and explain that it indicates the angle: the opening, or space, between the two sides. *What is the unit we will use to measure angles?* [Degrees]

4. Explain that students will measure different types of angles (acute, right, obtuse, and straight). *What type of angle is shown here?* Press *New Angle* several times. Access students' prior knowledge by asking them to name each type of angle and to define it. Write the definitions on chart paper after each new type of angle is encountered.

Type of Angle	Definition
Acute	Measures less than 90°
Right	Measures exactly 90°
Obtuse	Measures between 90° and 180°
Straight	Measures exactly 180°

5. Now refer to the protractor on the sketch. *What do you notice about this tool called a protractor?* Students may respond that it is half a circle (*semicircular*), it has a 180-degree scale, and the scale is named in intervals of 30 degrees. Model how to use the protractor to measure ∠*ABC*.

   • Drag the protractor so its center, the *origin*, is on point *B*, the vertex of the angle. To move the protractor, drag the origin.

   • The baseline of the protractor is likely to be aligned with side *BA* of ∠*ABC*. Rotate the protractor so the baseline aligns with side *BC*, and discuss why this position does not work. Rotate the protractor again to align the baseline with side *BA*. To rotate the protractor, drag any point on its circumference.

Let students know that when the measure of the angle Sketchpad displays is 0°, no angle measurement will be reported.

6. Ask students to name the type of angle shown and then use the protractor to estimate the angle measure. ***Between which two measures on the scale does side BC intersect the protractor?*** Introduce the language "It lies between $x°$ and $y°$." ***Is this an exact measurement?*** [No] Elicit the idea that, except when the computer displays an angle of 30°, 60°, 90°, 120°, 150°, or 180°, students can get a rough estimate of the angle measure using this protractor, but not a precise measurement. Press *Show Measurement* and explain that Sketchpad displays a more precise measurement.

A common student misconception is that changing an angle's side lengths changes the size of the angle.

7. Go to page "180 B." Discuss three differences in this model: the protractor's scale is marked for 10° increments; pressing *New Angle* displays angles in differing orientations; and dragging the endpoint of the slider changes the length of an angle's sides. Position the protractor by dragging and rotating, and then estimate the angle measurement. Press *Show Measurement*. Lengthen the angle's sides. Students should note that the angle measurement stays the same.

DEVELOP

Expect students at computers to spend about 30 minutes.

8. Assign students to computers and tell them where to locate **Introducing Protractors.gsp.** Distribute the worksheet. Tell students to work through step 7 and do the Explore More if they have time. Encourage students to ask their neighbors for help if they are having difficulty with Sketchpad.

9. Let pairs work at their own pace. As you circulate, check that students are taking turns using the protractors and are able to perform the following tasks.

On page "180 C," it is possible to drag the protractor by clicking on its circumference; however, this will create a point. Instead, drag point *O,* the origin.

• Drag the protractor so that the origin is on the angle's vertex. ***Where should you move the origin of the protractor? Which point is the vertex of the angle?***

• Rotate the protractor to align the baseline with a side of the angle. ***How should you position the protractor to measure the angle?***

• Drag the slider's endpoint to increase the length of the angle's sides. ***What can you do to see where this side of the angle intersects the protractor's scale?***

- Estimate angle measures using the language "between $x°$ and $y°$." *Between which two measures on the scale does the angle's other side intersect the protractor?*

- Verbalize what the marks on the protractor's scale represent. *Describe the scale on this protractor.*

- Classify angles and use the classification to check that estimates makes sense. *What type of angle is this? Is your estimate reasonable? Explain.*

- For page "180 C," realize that the scale on the protractor is more precise. *How is the scale of this protractor different?*

10. If students have time for the Explore More, they will measure an angle where neither side is aligned with the baseline on the protractor. (The sketch initially shows $\angle ABC$ with side BA at 60° and side BC at 120°.) Students will drag point A and/or point C to construct four more angles and measure them without aligning the sides of the angles with the baseline on the protractor.

SUMMARIZE

Project the sketch. Expect to spend about 10 minutes.

11. Gather the class. Students should have their worksheets with them. Begin the discussion by opening **Introducing Protractors.gsp.** Briefly show each of the following pages: "180 A," "180 B," and "180 C." *Today you used three different protractors to measure angles. Compare the scales on the protractors and explain how precise each one was.* Here are sample observations.

   The first protractor's scale had marks every 30 degrees. The second protractor had marks every 10 degrees. The last protractor had marks every 1 degree. The last protractor was the most precise.

   When we measured using the first protractor, our measures were given within a span of 30 degrees. The second protractor was better; that span was 10 degrees. For the last protractor, we could measure to the degree; it was the most precise tool.

12. Distribute physical protractors, straightedges, and pencils. Go to page "180 C." *Compare the protractor you are holding with this Sketchpad protractor. How are they alike? How are they different?* Students may give responses such as these, depending on the protractors in your class.

Both protractors have scales marking every degree, up to 180°.

The Sketchpad protractor has one scale. This protractor has two scales. It goes from 0° to 180° and 180° to 0°.

The Sketchpad protractor labels every 30 degrees: 0, 30, 60 . . . . The other protractor labels every 10 degrees: 0, 10, 20, 30 . . . .

In Sketchpad, the origin and the baseline are right on the edge. On this protractor they aren't at the edge of the protractor. And the origin of this protractor is a hole.

13. **What will you need to do differently when you use these protractors?** Have students share their ideas. The three important points to bring out are these.

    • Position the protractor carefully without using the bottom edge so the origin of the protractor is on the vertex of the angle and the baseline aligns with a side of the angle.

    • Read the correct scale.

    • Use a straightedge and pencil to make the sides of angles longer, if needed, so that they intersect the scale.

14. **How might thinking about the type of angle before you measure help?** [It will help students avoid reading the wrong scale; it is a check for the reasonableness of their answers.]

15. Have students draw an angle on the worksheet and try using the conventional protractor to measure its size. Then have students measure the size of an angle drawn by a neighbor. Discuss. **Did anyone have to extend the sides of the angle to see where they intersected the scale of the protractor? What else did you notice?**

16. If time permits, discuss the Explore More. Have students share their strategies for finding the angle measures.

EXTEND

1. For students who would benefit from more individualized preparation for using protractors, provide additional time for them to use the sketch.

2. Go to page "Extend." For the quadrilateral shown, have students name the type of angle at each vertex and then measure each angle using the protractor on screen. ***What is the sum of the angle measures?*** Now have students change the shape of the quadrilateral several times, and each time measure the size of the angles and find the sum of the measures. ***Is the sum of the angle measures the same?*** The sum of the angles of a quadrilateral is 360°. However, because the measurements students make using the protractor are not exact, students may frequently find the sum to be close to but not exactly 360°. Allow time for discussion. What are students' ideas about the precision of the measurements they made with the protractor? What are their ideas about the sum of the angles of a quadrilateral?

ANSWERS

3. Check students' work.

4. Check students' work.

5. Check students' work.

6. Answer will vary. Students should note that the scale is more precise. It measures to the nearest degree.

7. Check students' work.

8. The measure of angle *ABC* is 60°. Answers will vary. Students may say that they subtracted 60° from 120° to get 60° or that they simply counted the number of degrees in between the angle sides on the scale.

9. Check students' work.

Protractors

Name: _____

Use a protractor to measure acute, right, obtuse, and straight angles.

EXPLORE

1. Open **Introducing Protractors.gsp.** Go to page "180 A."

2. Practice moving and rotating the protractor.

   To move the protractor, drag its center.

   To rotate the protractor, drag any point on the circumference.

3. Now use the protractor to estimate the measure of ∠ABC.

   Drag the protractor so its origin is on point B, the vertex of the angle.

   Rotate the protractor so the baseline aligns with side BA.

   Use the protractor to estimate the angle measure.

   Side BC intersects the protractor between _____ and _____.

   Angle ABC is about _____°.

4. Now follow these steps. Fill in the table for four new angles.
   • Press *New Angle.* What type of angle appears?
   • Use the protractor to estimate the angle measure. (Take turns.)
   • Press *Show Measurement.* What is the measurement?
   • How close was your estimate?

Type of Angle	Estimate	Measurement	How Close Was Your Estimate?

5. Go to page "180 B." Fill in the table for four angles.

Type of Angle	Estimate	Measurement	How Close Was Your Estimate?

6. Go to page "180 C." What is different about the scale on this protractor?

_____

_____

7. Follow these steps. Fill in the table for six angles. Make some angles obtuse.

- Drag points *A* and *C* to change the angle.
- What type of angle did you make?
- Use the protractor to measure the angle.

Type of Angle	Protractor Measurement

EXPLORE MORE

8. Go to page "Explore More." Without moving the protractor, find the measure of ∠ABC. Explain how you figured it out.

_____

_____

9. Drag point *A* and/or point *C* to construct other angles. Do not align side *BA* with the baseline of the protractor. Measure at least four angles.

New York City Title I Elementary School Activities with The Geometer's Sketchpad
© 2012 Key Curriculum Press

Measure by Degrees: Types of Angles

INTRODUCE

Project the sketch for viewing by the class. Expect to spend about 10 minutes.

1. Explain, *Today you're going to use Sketchpad to explore angles and angle measures. An angle is sometimes defined as two rays that share an endpoint. Two segments with a common endpoint can also determine an angle. An angle can be named after three points or one. The angle measure is different from the angle.*

2. *I'll demonstrate how to construct and measure an angle, and then you'll construct and measure angles on your own.* Start with a new sketch and enlarge the document window so it fills most of the screen. As you demonstrate, make lines thick and labels large for visibility. First, model the angle construction in worksheet steps 1–5. Then model how to find an angle measure in worksheet step 8. Here are some tips.

   - In worksheet step 1, explain, *Angles are measured in degrees. For today's exploration we will want to measure in degrees to the nearest tenth.* Model how to choose **Edit | Preferences | Units** and set Angle Units to **degrees** and Angle Precision to **tenths**.

   To change a label, choose the **Text** tool and double-click the label. In the dialog box, type a new label and click **OK**.

   - In worksheet step 2, construct $\overrightarrow{AB}$. Model choosing the **Ray** tool. Labels for the points will automatically appear if you start with a new sketch and choose **Edit | Preferences | Text** and check **Show labels automatically: For all new points.** Model how to add or change labels using the **Text** tool.

   - Once the points are labeled correctly, click on the two points in this order: endpoint of the ray (point A) and the other point on the ray (point B). Explain that students will wait to label the points until the construction is complete; you are labeling now so that you can refer to the points as you demonstrate.

   - In worksheet step 3, use the **Ray** tool to construct $\overrightarrow{AC}$. *When endpoint A is selected, the point will be highlighted. When it is highlighted, click point A and drag to construct the new ray, the second side of the angle.* Make sure the third point is labeled C.

   - In worksheet step 4, model dragging the three points around with the **Arrow** tool. Emphasize the importance of this drag test to be sure that the two rays share a common endpoint.

   - Make sure students know how to use **Edit | Undo** if they make a mistake.

- Now model how to measure the size of the angle. ***The points of the angle must be selected in the correct order with the vertex as the second point.*** Select the three points, with point *A* second, and then choose **Measure | Angle**. Read the measure aloud, noting that the degrees are in tenths as desired. ***The*** **m** ***in front of the angle stands for "the measure of."***

- In worksheet step 8, after measuring the angle, drag point *C* to demonstrate that the angle measure updates automatically as the size of the angle changes.

3. If you want students to save their work, demonstrate choosing **File | Save As,** and let them know how to name and where to save their sketches.

DEVELOP

Expect students at computers to spend about 20 minutes.

4. Assign students to computers. Distribute the worksheet. Tell students to work through step 21 and do the Explore More if they have time. Encourage students to ask their neighbors for help if they are having difficulty with the construction.

5. Let pairs work at their own pace. As you circulate, here are some things to notice.

- For worksheet steps 6 and 7, look to be sure students understand what the vertex of the angle is. Restate, if needed, that the vertex is the point where the two sides of the angle meet. Introduce or review naming an angle using the angle symbol and the label of the vertex.

- In worksheet step 10, students should notice that the angle measure increases and then begins to decrease once it passes 180°. Sketchpad acts like most protractors, so it does not measure reflex angles.

- For step 12, students may not describe the rays as perpendicular; instead they may say the angle looks like "a corner." Encourage students to use geometric terms in their descriptions.

- Since worksheet step 16 was not modeled, some students may not successfully construct the circle centered at point *A*. Tell students to be sure point *A* is highlighted before they click and drag to construct the circle. Mention the drag test to check that point *A* remains the center of the circle.

• For step 17, check that students select only the circle and point *C*.

• For step 21, tell students that they can look back to steps 2–5 for help, but they should choose the **Segment** tool, not the **Ray** tool, to construct the angle this time. Notice students who are unsure whether the size of the angle depends on the lengths of the angle's sides. Are they able to convince themselves that this is not so by lengthening the sides while decreasing the angle measure, or by shortening the sides while increasing the angle measure?

• The Explore More helps students develop visualization and estimation skills. They make angles of different sizes, classify each angle, and estimate the angle measure before measuring to get immediate feedback.

SUMMARIZE

Project the sketch. Expect to spend about 15 minutes.

6. Gather the class. Students should have their worksheets. Open **Measure by Degrees Present.gsp** and use page "Angle Measure" as needed. Begin the discussion by checking that students understand how an angle is formed. *What did you use to construct the sides of your angle in steps 2–5? In step 21? Based on your constructions, how would you define angle?* Write, "An angle is" on chart paper or other display and help the class create a clear definition. Here is a sample definition: An angle is a figure formed by two rays, or segments, that share a common endpoint called the *vertex*. Review the terms *ray, segments, endpoint,* and *vertex* as needed.

7. Review worksheet steps 6 and 7. *What are the different ways to name an angle?* Write "An angle can be named" on the display. Students should reply that an angle can be named by its vertex, such as $\angle A$ for the first construction, or by three points: one point on one side of the angle, the point at the vertex, and one point on the other side, such as $\angle BAC$ or $\angle CAB$. Be sure students understand that the vertex is always the second point when naming an angle using three points. Introduce the angle symbol if students are not familiar with it.

8. Discuss worksheet steps 11–15 with the class. Have students define the different types of angles: right, straight, acute, and obtuse. Add the definitions to the chart paper. (You don't need to introduce *reflex angle* at this time.) Then ask students to classify the different angles they see in their environment. Students may suggest the following ideas.

Right angles are found in the corners of windows, doors, walls, desks, and my worksheet.

Where the chair leg meets the seat is a right angle.

The hands of the clock make an acute angle when the time is 10:00. The hands make an obtuse angle when the time is 10:15.

The roof line of the school forms an obtuse angle.

The bottom of the white board is a straight angle.

9. Discuss worksheet step 20. Have students share how they arrived at their answers. ***If a complete revolution forms a circle, how many degrees are in a circle?*** Talk about how many degrees are in one-fourth and one-half revolutions. ***What angles do these represent?*** [Right and straight]

10. Have students discuss their responses to worksheet step 21; invite a volunteer to the computer to demonstrate. Be sure students understand that the length of an angle's sides is not related to the measure of its angle, a key misconception that many students have.

11. If students had time for the Explore More, ask them to share their findings.

ANSWERS

6. The vertex of the angle is point *A*. The name of the angle using just the vertex is $\angle A$.

7. $\angle BAC$, $\angle CAB$

10. The smallest angle measure is 0°. The greatest is 180°.

11. An angle with measure 0° is a single ray.

12. An angle with measure 90° has perpendicular sides.

13. An angle with measure 180° is a straight line.

14. Drawing shows acute angle.

15. Drawing shows obtuse angle.

20. The greatest angle measure would be 360°.

21. If an angle has segments for sides, it is possible to lengthen the sides while actually decreasing the angle measure.

23. Angle sketches and measures should match.

Measure by Degrees

Name: _____

In this activity you'll explore angles and angle measures.

CONSTRUCT

1. In a new sketch, choose **Edit | Preferences** and go to the Units panel.
   Set Angle Units to **degrees** and Angle Precision to **tenths**. Click **OK**.

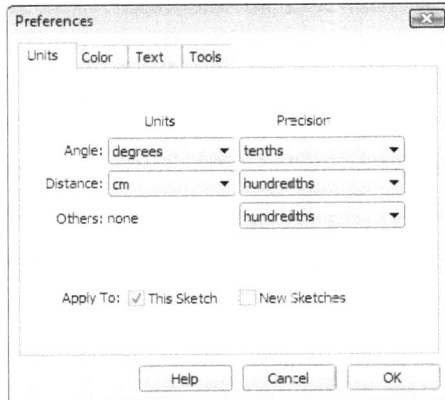

Preferences			
Units	Color	Text	Tools

	Units	Precision
Angle:	degrees ▼	tenths ▼
Distance:	cm ▼	hundredths ▼
Others: none		hundredths ▼

   Apply To: ☑ This Sketch ☐ New Sketches

   Help   Cancel   OK

2. Construct $\overrightarrow{AB}$.

3. Now construct $\overrightarrow{AC}$ with the same endpoint A.

4. Drag each of the three points to make sure the two rays share a common endpoint (point A).

   If you need to, choose **Edit | Undo** to back up one or more steps.

5. If necessary, use the **Text** tool to display the point labels. Change them to match the figure.

   To change a label, double-click the label. In the dialog box, type a new label and click **OK**.

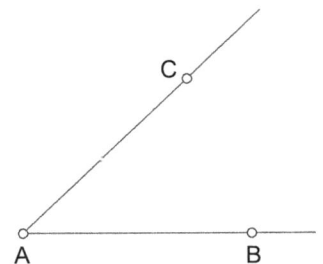

EXPLORE

6. Two rays with a common endpoint form an angle. The common endpoint is called the *vertex* of the angle. Sometimes angles are named just by their vertex. What is the vertex of the angle you just made? _____

   What is the name of your angle using just the vertex? ∠ _____

Measure by Degrees

continued

7. Angles can also be named after three points: a point on one side, the vertex, and a point on the other side. The vertex is always named second. What are two other possible names for the angle you just made?

   ∠ _____ and ∠ _____

8. Select, in order, points *B*, *A*, and *C*. Choose **Measure | Angle.** When you select an angle in Sketchpad, the vertex must be your second selection.

9. Drag point *B* or point *C* and observe how the angle measure changes.

10. What is the smallest possible angle measure? _____

    What is the greatest possible angle measure? _____

11. Drag a point on your angle until the angle's measure is as close to 0° as possible. Describe this angle.

12. Drag a point on your angle until the angle's measure is as close to 90° as possible. Describe this angle.

13. Drag a point on your angle until the angle's measure is as close to 180° as possible. Describe this angle.

14. An acute angle has a measure between 0° and 90°. Drag a point on your angle to make it acute. Sketch an example of an acute angle.

15. An obtuse angle has a measure between 90° and 180°. Drag a point on your angle to make it obtuse. Sketch an example of an obtuse angle.

16. Construct a circle centered at point *A* but not attached to any other points in your sketch.

New York City Title I Elementary School Activities with The Geometer's Sketchpad
© 2012 Key Curriculum Press

Measure by Degrees

17. Select the circle and point C.

    Then choose **Edit | Merge Point to Circle**.
    Point C will attach itself to the circle.

    m∠BAC = 52.6°

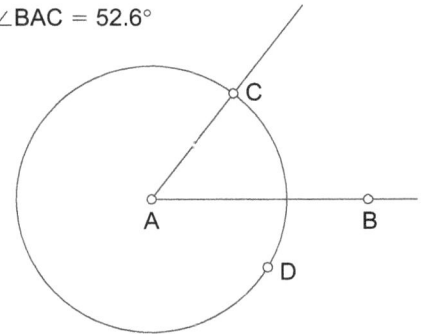

18. Now you will make an action button to animate
    point C around the circle.

    Select point C and choose
    Edit | Action Buttons | Animation.

    Set Speed to **slow** and click **OK**.

19. Press *Animate Point*. Watch the angle measurement. Stop the animation by
    pressing the button again.

20. Because of the way Sketchpad measures angles, when one angle side rotates
    by more than one-half a revolution from the other angle side, the angle
    measure starts to decrease. Suppose the angle measure kept increasing until
    the angle side had completed one whole revolution.

    What would be the greatest angle measure? _____

21. Construct a new angle using the **Segment** tool.

    Does the measure of the angle depend on the lengths of its sides? Explore this
    question in Sketchpad. Then give your answer and explain your reasoning.

EXPLORE MORE

22. Get ready to try estimating the sizes of angles.

    Using the angle you made in step 21, measure the angle.

    With the measure selected, choose **Edit | Action Buttons | Hide/Show**.
    Now you can press this button when you want to hide or show the
    angle measurement.

23. Drag the endpoints of the segments to make new angles. Estimate the angle sizes and then measure them. Record your work in the table. Remember to hide the angle measurement before you estimate.

Type of Angle	Estimate	Measure

New York City Title I Elementary School Activities with The Geometer's Sketchpad
© 2012 Key Curriculum Press

Straight Ahead: Segments, Lines, and Rays   For GSP5   ACTIVITY NOTES

INTRODUCE

Project the sketch for viewing by the class. Expect to spend about 10 minutes.

1. Open Sketchpad and enlarge the new sketch so it fills most of the screen. Explain, *Today you're going to draw objects that are straight. There are several tools in Sketchpad that let you draw straight objects. They're called* Straightedge tools. Write *straightedge* on the board. Explain that a straightedge is a ruler without marks; it can be used to draw straight objects, but not for measuring.

For now, ask the class to agree to call everything on screen "an object." Wait to give names until students have had time to explore the tools and objects on their own.

2. Point out the **Straightedge** tool icon in the Toolbox. Press and hold on the icon to show the menu that pops out. Without naming it, select the **Segment** tool. Click in two places to draw a segment. Repeat several times.

3. Select the **Arrow** tool. *After we construct an object, we can change it.* Select the endpoint of a segment, note that it is highlighted, and drag the endpoint to change the length and direction of the segment.

   Using the **Arrow** tool, select the segment itself (not the endpoints), note that it is now highlighted, and drag the segment. Have students observe that the segment moves but does not change its length or direction.

4. *We can also get rid of an object if we don't want it.* Using the **Arrow** tool, select a segment *and* its endpoints and choose **Edit | Clear Objects.**

5. Briefly introduce the other two **Straightedge** tools. Referring to them only as "objects," draw several rays and lines. Have students observe that these objects are also drawn by clicking in two places, creating two points. Using the **Arrow** tool, drag the points.

6. Model how to change the appearance of an object. Making sure the object is selected, choose **Display | Line Style | Thick,** observe the result, and then choose **Display | Line Style | Dashed.** Choose **Display | Color** and choose a color. Repeat to change the color a few times.

7. *Now you'll explore the Straightedge tools on your own. Try to find out more about each tool. We'll talk about what you discover.*

   If you will have students print their sketches, model how to choose **File | Print Preview,** and in the dialog box that appears, set the image to fit on one page and then click **Print.**

If you want students to save their work, demonstrate choosing **File | Save As,** and let them know how to name and where to save their files.

DEVELOP

Expect students at computers to spend about 35 minutes.

8. Assign students to computers, and tell them where to locate **Straight Ahead.gsp.** Distribute the worksheet. Tell students to work through steps 1–7 and do the Explore More if they have time. Encourage students to ask a neighbor for help if they have questions about using Sketchpad.

9. Let pairs work at their own pace. As you circulate, here are some things to notice.

   • In worksheet step 1, encourage students to fill their screens with lots of segments, lines, and rays. This is a good opportunity for students to begin associating each tool in the Toolbox with the type of object it creates.

   • In worksheet step 4, students drag each type of object: a segment, a ray, and a line. ***What have you noticed as you've dragged the different objects?*** Here are sample student responses.

     I tried to find where this object [a line] ended, but no matter how much I dragged it, it seemed like it went on forever.

     This object [a ray] only has one end, just like the picture in the Toolbox. I couldn't find another end no matter how much I dragged.

If students are familiar with the terms *segment, line,* and *ray,* they may introduce the terms into conversations.

   • In worksheet step 4, some students may think that lines *do* have a finite length because they are framed by the edges of the sketch window. Explain that the sketch window shows only a portion of the line. If the computer screen were a lot bigger, more of the line would be visible.

   • In worksheet step 5, listen as students compare the **Straightedge** tools. You may hear comments such as these.

     We can drag all of the objects and change their widths and colors.

     All the tools make straight lines. None of the tools can make curves.

     Some of our objects have ends. This one has one end, this one has two ends, but this one doesn't have any.

All the objects have two points. You can always drag the points to change direction.

- In worksheet step 6, to draw a segment or ray that begins on the circle and is attached to it, students need to (1) select a tool and move it to the desired place on the circle's circumference, and (2) watch to see that the circumference is highlighted before clicking.

- In worksheet step 7, students will have fun finding and fixing the differences between two stick figures. To fix Girl B's left arm, students will need to delete the ray and draw a new arm using the **Segment** tool. Give them time to think this through and consult each other; don't provide a hint right away.

10. If students will save their work, remind them where to save it now. If you want students to print their sketches, remind them how to do so.

SUMMARIZE

Project a new sketch. Expect to spend about 15 minutes.

11. Gather the class. Students should have their worksheets with them. In a new sketch, construct a segment. *What can you say about this object? How is the tool's picture, or icon, in the Toolbox similar to what it constructs?* Here are sample responses.

This tool makes a straight line that has a point at each end. It looks just like its picture.

It's like a straight path with brick walls at both ends. If a robot was walking along the path, it couldn't continue past the walls.

The object has a beginning and an end.

12. Construct a ray. *What's different about this object? How is the tool's picture, or icon, similar to what it constructs?*

The picture shows an arrow at one end and a point at the other. The object starts at one point and goes out forever in a straight path. I think the arrow means the straight line keeps going.

This object has two points, too, but the line stops at one and continues going past the other.

Well, if the robot walked along this straight path, it would hit one brick wall, but it would be able to walk in the other direction and continue without stopping.

13. Construct a line. *What's different about this object? How is the tool's picture, or icon, similar to what it constructs?*

   The straight line goes in both directions and never ends. The picture shows this by putting arrows at both ends.

   This one has two points, and a straight line goes through both points and continues forever in both directions.

   The robot can walk in a straight path in both directions without stopping. No brick walls!

14. *Should we call all of these objects* lines? Take responses. Explain that in mathematics, there are different names for the different objects. Introduce each term—*segment, line,* and *ray*— by writing and saying its name as you point to the corresponding objects on the sketch. Also introduce the term *endpoint*. A segment has two endpoints, a ray has one endpoint, and a line has no endpoints.

   Suppose you were writing a math dictionary. How would you define a line? Facilitate discussion for each term, helping the class clarify the definitions. The sample definitions here are mathematically correct.

   • A line is a straight path that extends without end (infinitely) in two directions.

   • A segment is a part of a line between, and including, two endpoints.

   • A ray is part of a line that has one endpoint. It extends without end (infinitely) in one direction.

15. In a new sketch, draw a segment, line, and ray. Using the **Text** tool, label the six points created.

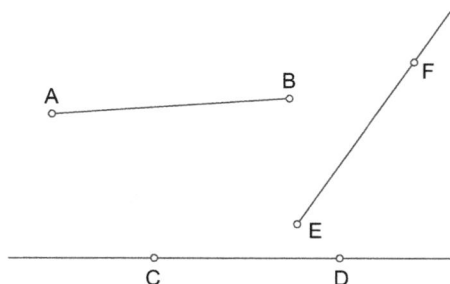

Introduce the notation used when writing the name of a segment, line, or ray.

- A segment is named by its two endpoints. The notation for writing segment *AB* is $\overline{AB}$, where points *A* and *B* are the endpoints. Students should see the connection between the notation and the **Segment** tool's icon.

- A line is named by any two points on the line. The notation for writing line *CD* is $\overleftrightarrow{CD}$. Students should see the connection between the notation and the **Line** tool's icon.

- A ray is named by its endpoint and another point on the ray. The notation for writing ray *EF* is $\overrightarrow{EF}$, where point *E* is the endpoint and point *F* is on the ray. Discuss the similarity between the notation and the **Ray** tool's icon.

16. If time permits, discuss the Explore More. Ask students to point out the different objects (segment, ray, line, point) they used.

EXTEND

1. To introduce the **Length** command, draw and select a segment (but not its endpoints) and choose **Measure | Length.** Students will observe the measurement displayed. Try to do the same for a line. Students will observe that the **Length** command is grayed out when you go to choose it. *Is it possible to measure the length of a line?* Students should respond that because a line continues infinitely in both directions, it would be impossible to measure its length. Have students tell whether they think Sketchpad will gray out the **Length** command if you try to measure a ray. [It will because it is impossible to measure the length of a ray.]

2. Have students name the Sketchpad tool—**Segment, Line,** or **Ray**—they would use to draw the top edge of a door [**Segment**], a number line [**Line**], and a flashlight beam [**Ray**]. *Think of some other things and tell which tool you would use to draw them.*

ANSWERS

5. Answers will vary. The tools are similar in that they all construct straight objects with two points. The tools differ in that

   • one tool constructs an object that has two endpoints;

   • one tool constructs an object that has only one endpoint and that extends infinitely in one direction; and

   • another tool constructs an object with no endpoints, extending infinitely in both directions.

6. Check students' work. Students will likely use a combination of segments and rays to complete a picture of the sun.

7. Students should have made these changes to Girl B: Add the missing hair; replace the girl's left arm with a segment; drag the feet to change their shape; show the bottom of the dress by drawing a segment or by dragging the endpoints of the belt down; delete the belt; and drag the girl's right elbow to change its position.

Straight Ahead

For GSP5 Name:

Explore the objects you can construct with the **Straightedge** tools.

EXPLORE

1. Open **Straight Ahead.gsp.** Go to page "Explore."
   Construct objects using each of the three **Straightedge** tools.

2. Change the line width of some of your objects. Select an object, choose **Display | Line Style,** and choose a width.

3. Change the color of some of your objects. Select an object, choose **Display | Color,** and choose a color.

4. Drag the straight objects. Drag the points. Watch what happens and be ready to talk about what you see.

5. Compare the objects you can make using the **Straightedge** tools.

   How are they alike?

   _____

   _____

   _____

   How are they different?

   _____

   _____

   _____

   _____

6. Go to page "Sun."
   Finish the picture any way you want.
   Change line widths and colors.

7. Go to page "Stick Figures."

   Change Girl B so she matches Girl A.

   If you need to delete an object, select it and choose **Edit | Clear.**

EXPLORE MORE

8. Go to page "Make Your Own 1."

   Construct your own stick figure.

   Have your partner try to make one just like it.

9. Go to page "Make Your Own 2."

   Construct your own picture.

   Use each straightedge tool at least once.

   Use different widths and colors.

Mondrian in Motion:
Parallel and Perpendicular Lines

For
GSP5

ACTIVITY NOTES

INTRODUCE

Project the sketch for viewing by the class. Expect to spend about 15 minutes.

1. Display images you have collected of Piet Mondrian's work and invite students to share what they observe, including observations about the geometric objects. Some students are likely to use the terms *parallel* and *perpendicular*. To assess students' understanding of these terms and to get them thinking about them, ask what these words mean. Don't offer definitions at this time.

2. Explain that students will explore these types of lines as they create their own works like Mondrian's, using Sketchpad. Open **Mondrian in Motion.gsp.** Using the worksheet and the procedure listed below, model what students will need to know to work on their own. (Students will benefit from doing some steps without having seen them first.)

   • Model steps 1–11, omitting step 4. Call special attention to the process in step 9 for constructing a polygon. (Note that you are also constructing vertices at the intersections of the lines.)

   • Show how to choose **Edit | Undo** to back up a step if you want to do something differently.

   • Choose the **Point** tool and then model locating the menu commands **Edit | Select All Points** and **Display | Animate Points.** Explain that students will use these commands when they have completed their artwork.

   • Press *Show Point A* to reveal a point *A* that lies on the blue line. Explain that students do not need to show point *A* at the beginning of their work. It will be used in the last step.

   • If you want students to save their Sketchpad work, demonstrate choosing **File | Save As,** and let them know where to save.

DEVELOP

Expect students at computers to spend about 30 minutes.

3. Assign students to computers and tell them where to locate **Mondrian in Motion.gsp.**

4. Distribute the worksheet and tell students to work through steps 1–15. Encourage students to ask their neighbors for help if they have questions about using Sketchpad.

New York City Title I Elementary School Activities with The Geometer's Sketchpad
© 2012 Key Curriculum Press

215

5. Let students work at their own pace. As you circulate, here are some things to be alert to as students work through the worksheet steps.

   • In steps 3 and 5, students may prefer to construct their parallel and perpendicular lines one at a time by selecting the blue line and *a single point* and then choosing **Construct | Parallel Line** (or **Perpendicular Line**).

   • Steps 3–6 prompt students to make the connection between their constructed lines and their ideas about the meanings of *perpendicular* and *parallel.*

   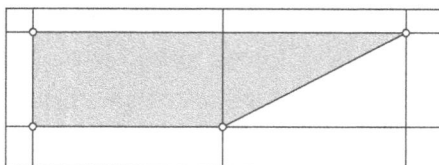

   • In steps 9 and 10, students are to construct and color in rectangles. If students forget that they are to construct rectangles (which include squares), they may construct a quadrilateral such as the one shown here. If you see this happening, bring students' attention back to the directions and to the shapes used in Mondrian's work.

   • In step 10, you may suggest that students construct and color as many rectangles as they choose, if time allows.

   • Once students have colored in shapes, they animate the points, which move freely in the plane. Students should notice that as the points move, the lines maintain their relationships, parallel or perpendicular, to the blue line.

   • In step 15, having students drag point *A* is intended to prompt them to consider what happens to their constructions when the blue line is no longer vertical. Students may be surprised to see that all the lines rotate along with the blue line and that the colored shapes remain rectangles. Note which students are comfortable with this and which continue to drag point *A,* unconvinced that this will be true no matter how they drag the blue line.

If you have a flash drive, use it to collect students' sketches for displaying on the shared computer. You can also share the artwork by holding a gallery walk.

6. Make sure students have "signed" their artwork. Have them save their sketches for the class discussion to follow, preferably in a place that will be accessible to the computer connected to the large-screen display.

7. If students will print their work, give the following instructions now.

   • Using the **Arrow tool**, select point A and the blue line. Now choose **Display | Hide Objects.**

   • Choose **File | Print Preview.**

   • In the dialog box that appears, set the image to appear on one page and then click **Print.**

SUMMARIZE

Project the sketch on a large-screen display for viewing by the class. Expect to spend about 10 minutes.

8. Bring the class together to discuss what they have experienced and observed. Invite students to reflect on their ideas about parallel or perpendicular lines. You may wish to introduce questions such as these into the discussion.

   What happens to the lines as you drag point A? Why? Students should realize that the lines remain parallel and perpendicular, even though their orientation, locations, and distances from one another change.

   Can you drag any of your points to make a line that is not parallel or perpendicular to any other line? Why? Students should begin to see that the lines always remain parallel or perpendicular to the blue line because they were constructed that way. The relationship the lines have to the blue line cannot change.

9. Continue the discussion, inviting students to share their thinking about the quadrilaterals in their work. Some questions to introduce follow.

It's also possible in Sketchpad for one or both dimensions of a rectangle to go to zero, in which case the rectangle disappears.

   What happened to the quadrilaterals as you dragged point A? Students should realize that the quadrilaterals remain rectangular as their orientation, locations, sizes, and proportions change. (Students are also likely to mention that when two rectangles overlap, their intersection may be a new color.)

   What shapes are the colored figures in your artwork? Students may at first describe the shapes as rectangles and squares.

Can you use the word quadrilateral *to describe any of the shapes you see?* Elicit the idea that all the shapes are four-sided polygons, so all the shapes are quadrilaterals.

What else can you say about all these shapes? Elicit the fact that the *opposite* sides of these quadrilaterals are parallel. Quadrilaterals whose opposite sides are parallel are called *parallelograms*, and specific parallelograms that the students see are called rectangles and squares. Students may also note that all the angles in these parallelograms are right angles.

What is true about the parallelograms called rectangles or squares? Parallelograms whose adjacent sides are perpendicular are rectangles and squares. All the angles are right angles. You may want to ask students to name and draw some parallelograms that do not have four right angles.

ANSWERS

4. Answers will vary. Possible answer: The lines are always parallel to the blue line. They never intersect the blue line, although they can coincide with it or move to the other side of it.

6. Answers will vary. Possible answer: These lines are perpendicular to the blue line. They cross the blue line at right angles.

7. Answers will vary depending on students' understanding of classes of shapes. It's possible to identify rectangles and squares, all of which can also be identified as quadrilaterals and parallelograms.

8. No.

13. Answers will vary. Possible answer: The lines stay parallel or perpendicular to the blue line. The shapes are still rectangles but they stretch and shrink.

15. Answers will vary. Possible answer: The lines remain parallel and perpendicular, and the quadrilaterals remain rectangles, even though the orientation, locations, sizes, proportions, and distances may change.

Mondrian in Motion

Like the painter Mondrian, you will make a work of art using lines and shapes.

CONSTRUCT AND EXPLORE

1. Open **Mondrian in Motion.gsp.** Go to page "Blue Line."
   Make the window as arge as you can.

2. To the right of the blue line, construct five points.

3. Select the points and the blue line.
   Choose **Construct | Parallel Lines.**

4. Drag the new lines, one at a time. How do they behave?

   _____

5. Select the points and the blue line again.
   Choose **Construct | Perpendicular Lines.**

6. Drag the new lines. What is their relationship to the blue line?

   _____

7. What shapes are made by the lines?

   _____

8. Try to drag any line sc that it is neither parallel nor perpendicular
   to the other lines. Can you do it? _____

9. Choose a rectangle in your picture to color in. Going around the
   rectangle, click the intersections of the lines to create four vertices
   and a rectangle. Click the first vertex again to complete the
   rectangle.

   With the rectangle selected, choose **Display | Color** and choose
   a color.

10. Repeat step 9 to color in four more rectangles.

A 11. Sign your name to your work.

• 12. Now you will animate your picture.
    Choose the **Point** tool.
    Choose **Edit | Select All Points.**
    Choose **Display | Animate Points.**

13. Describe how the shapes behave. _____

_____

_____

14. To stop the animation, press the Stop button on the Motion Controller.
    You can also choose **Display | Stop Animation.**

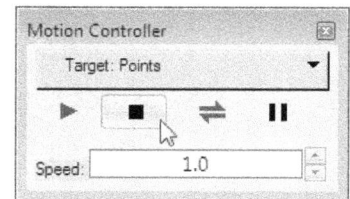

15. Press *Show Point A.* Slowly drag point *A.*
    How do the lines and shapes behave? _____

_____

_____

Perimeter Formulas: Algebraic Notation

For GSP5 ACTIVITY NOTES

INTRODUCE

Project the sketch on a large-screen display for viewing by the class. Expect to spend about 10 minutes.

1. Open **Perimeter Formulas.gsp.** Go to page "Rectangle." Explain, *In this activity, you will be exploring ways to find the perimeter of a rectangle. I'll introduce the Sketchpad model, and then you'll work on your own.* Introduce the model and any Sketchpad skills new to your students.

   • On page "Rectangle," have students describe how the rectangle behaves as you drag each vertex of the rectangle. [Dragging point *B* changes the lengths of both side *AB* and side *CD*. Dragging point *D* changes the lengths of both side *DA* and side *BC*. Dragging point *C* changes the lengths of all sides. Dragging point *A* translates the rectangle without changing its dimensions.]

   • Sketchpad's length measurements include an "m" that stands for *measure of.* On page "Measure," model measuring the length of a side of the rectangle and reading the displayed measurement.

   • On page "Calculate," students will measure in a way different from the way they do on page "Measure": They will select two vertices and choose **Measure | Distance.** Model measuring distance and then explain that the reason for doing this is to avoid having the symbol "m" displayed in the sketch. This will make the calculations students insert in the Calculator easier to read and refer to.

   • Students will discover the power of using this technique when they drag the rectangle and watch the displayed calculation update automatically.

   • Students will use Sketchpad's Calculator to create a calculation of the rectangle's perimeter. Unlike what they do when using a handheld calculator, students will not press buttons on the Calculator's keypad to insert the numbers for the value of a measurement. Instead, they will click directly on a measurement in the sketch to insert it into their calculation. Model clicking one of the displayed measurements to enter it into the Calculator. Also, point out the parentheses, addition, and multiplication buttons in the Calculator's keypad.

2. If you want students to save their Sketchpad work, demonstrate choosing **File | Save As,** and let them know where to save.

DEVELOP

Expect students at computers to spend about 25 minutes.

3. Assign students to computers and tell them where to locate **Perimeter Formulas.gsp.** Distribute the worksheet. Tell students to work through steps 1–11 and do the Explore More if they have time. Encourage students to ask a neighbor for help if they have questions about using Sketchpad.

4. Let pairs work at their own pace. As you circulate, here are some things to notice.

   • In worksheet step 2, students should spend time dragging the vertices of the rectangle and becoming familiar with how it behaves.

   • Students' responses to worksheet step 4 will help you assess their understanding that the opposite sides of a rectangle have the same length. If some students seem unsure of this, spend time with them carrying out step 5, observing the behavior of the sketch and talking about this property of rectangles. Students need this understanding to make sense of this activity.

   • In worksheet step 5, students should observe that the displayed measurements update automatically. This simple observation is important because it will help students understand how their calculations update automatically when they carry out step 8.

   • In worksheet step 7, encourage students to ask a neighbor for help if they have questions about using Sketchpad's Calculator.

   • In worksheet step 9, students discover the power of inserting measurements in the sketch into the Calculator. When they drag the vertices of the rectangle, they observe that the displayed calculation updates automatically.

   • Worksheet step 10 asks students to use multiplication as well as addition to calculate the perimeter. Some students may have done this already in step 7; ask them to create another calculation now. If some students need help, ask, *Is there a way to find the rectangle's perimeter without adding the lengths of all four sides? Can you use two sides only?* Here are some of the calculations students might enter.

$$2 * AB + 2 * BC \qquad 2 * (CD + AD)$$

$$2 * BC + 2 * AB \qquad (CD + AD) * 2$$

Students may have a correct idea, but not enter their calculation correctly into the Calculator. For example, some may enter "$CD + DA * 2$" because they want to add two adjacent sides of the rectangle and then double the result. Because the Calculator multiplies DA by 2 before adding CD, it will not compute what students expect. This is a good opportunity to review the role of parentheses and the order of operations.

5. Students now create perimeter calculations for a square and for an equilateral triangle. There are two approaches: add the lengths of all sides, or find the length of one side and multiply by the number of sides.

SUMMARIZE

Project the sketch. Expect to spend about 25 minutes.

6. Bring the class together. Have students share their methods for calculating perimeter. As they do, record them on the board. You might make three columns of related expressions as you record—without explaining your reasoning to students.

$AB + BC + CD + DA$	$2 * AB + 2 * BC$	$2 * (CD + AD)$
$AB + CD + BC + DA$	$2 * BC + 2 * AB$	$2 * (AD + CD)$
$AB + AB + BC + BC$	$(CD + AD) * 2$	

7. **Why do these different approaches give the same perimeter?** Some sample student responses are presented here.

It doesn't matter what order you add up the four side lengths in. As long as every side is included, the answer will be the same.

Opposite sides of the rectangle have the same length. So, multiplying AB by 2 is the same as adding AB and CD. And multiplying BC by 2 is the same as adding BC and DA. So, the calculation 2 * AB + 2 * BC will be the same as the calculation AB + CD + BC + DA.

It doesn't matter which long side and which short side you add before you multiply by 2, because both long sides are the same length and both short sides are the same length.

All the ways we calculated end up finding the total length of two long sides and two short sides.

8. ***How would you tell someone else, using words only, how to calculate the perimeter of any rectangle?*** Record students' descriptions on the board or chart paper.

   Add the lengths of all four sides.

   Add the length of a long side and the length of a short side. Multiply that by 2.

   Double the length, and double the width. Add the result.

9. Focus on the method "add all four side lengths." ***Use the words*** **length** ***and*** **width** ***and any operation signs to write statements representing your ways of computing the perimeter. Start your statements like this:***

   Perimeter =

   Invite volunteers to write statements. Some examples are these.

   Perimeter = length + width + length + width

   Perimeter = 2 × length + 2 × width

10. ***I don't want to write out all of these words. It takes too long! Mathematicians use letters in place of words. What letter shall we use to represent the length of the rectangle?*** Any letter is a possibility. Using l for length makes sense because it makes it easy to remember what l stands for. Similarly, w can represent width and P can represent perimeter. Choose one statement and write it using letters. For example: $P = l + w + l + w$.

    What do your other methods look like written this way? Record responses in a table such as this one.

Add all four side lengths.	$P = l + w + l + w$
Double a long side, double a short side, and add together.	$P = 2 × l + 2 × w$
Add a long and a short side, and double the result.	$P = (l + w) × 2$

11. Tell the class that the methods they have developed for finding the perimeter of a rectangle are called *formulas*. Give a definition: A formula is a general rule that describes a calculation.

12. If students had time to calculate the perimeter of the square and the equilateral triangle, ask, *How were your ways like your calculations for a rectangle? How were they different? What formulas would you write for finding the perimeter of a square? For an equilateral triangle?*

ANSWERS

3. Answers will vary.

4. Answers will vary. Students should give the length of side *AB* for the length of side *CD*, and the length of side *BC* for the length of side *DA* because opposite sides of a rectangle have the same length.

5. The measurements update as point *C* is dragged.

8. Possible answers are those listed in Activity Notes step 6.

9. The answer is yes if the calculation is correct.

10. Possible answers are listed in step 6 of the Activity Notes.

11. The answer is yes if the calculation is correct.

12. Calculation statements will vary. The possible approaches include finding the lengths of all four sides and adding them, and finding the length of one side and multiplying by 4.

13. Calculation statements will vary. The possible approaches include finding the lengths of all three sides and adding them, and finding the length of one side and multiplying by 3.

Perimeter Calculations

For GSP5 Name:

Create calculations for the perimeter of a rectangle.

EXPLORE

1. Open **Perimeter Formulas.gsp.** Go to page "Rectangle."

2. Drag point C. Try dragging the other points.
   Get to know how rectangle *ABCD* behaves.

3. Go to page "Measure."
   Select side *AB*. Choose **Measure | Length.**
   Find the length of side *BC* too. Record both lengths.

   _____

4. Without measuring, what do you know about the lengths of sides
   CD and *DA*?

   _____

5. What happens to the measurements as you drag point *C*?

   _____

6. Go to page "Calculate."
   Select point *A,* then point *B.* Choose **Measure | Distance.**
   Select only point *B,* and then point *C.* Measure the distance.
   Select only point *C,* and then point *D.* Measure the distance.
   Select only point *D,* and then point *A.* Measure the distance.
   If you make a mistake, choose **Edit | Undo** to back up.

7. Choose **Number | Calculate.**
   Calculate the perimeter of the rectangle.
   Don't key measurements into the Calculator!
   Do click on the measurements on screen.
   When you are done, click **OK.**

8. Write your calculation as it looks in the sketch.

   _____

9. Drag point C of your rectangle to change its perimeter.
   Look at the new measurements.
   Does your perimeter calculation work for new measurements? _____

10. Use the Calculator to find the perimeter of *ABCD* again.
    This time, use the Calculator's multiplication key (the * symbol)
    along with the addition key.
    Write your calculation as it looks in the sketch.

    _____

11. Drag point C. Look at the new measurements.
    Does your perimeter calculation work for new measurements? _____

EXPLORE MORE

12. Go to page "Square."
    Use the Calculator to find the perimeter in more than one way.
    Drag the shape to check your way of calculating.
    Record each way.

    _____

    _____

13. Go to page "Triangle" and do the same thing.
    Record the calculations.

    _____

    _____

Pool Border: Equivalent Expressions

For GSP5 ACTIVITY NOTES

INTRODUCE

Project the sketch for viewing by the class. Expect to spend about 15 minutes.

1. Open **Pool Border.gsp** and go to page "Problem." Enlarge the document window so it fills most of the screen. Ask for a volunteer to read it aloud. Make sure everyone understands the problem.

2. Say, *Let's start with a 10-foot pool. What's a good problem-solving technique for getting started?* Most students will want to draw a diagram, if they haven't already done so. Have a student volunteer draw a diagram on the board.

3. *Can we solve the problem?* Some students may count tiles. Others may offer other answers, using multiplication and addition or subtraction. For each case, ask students to draw a diagram on the display with the regions shaded to illustrate their reasoning. Also work with the class to write each suggestion as an arithmetic expression involving the number 10. Encourage creative thinking. These are some of the many possibilities.

   - $10 + 10 + 10 + 10 + 4$
   - $4 \times 10 + 4$
   - $4 \times (10 + 1)$
   - $4 \times (10 + 2) - 4$
   - $2 \times (10 + (10 + 2))$
   - $12 \times 12 - 10 \times 10$

   If students omit parentheses and get an answer other than 44, take advantage of the situation to review standard order of operations. For example, if they write $4 \times 10 + 1$, they will get 41, because multiplication precedes addition; only with parentheses to make $4 \times (10 + 1)$ will the addition be done first.

4. *If the pool were a different size, we'd use a number other than 10. Would the expressions still give the same result? For example, if the side of the pool were any length x, would $4 \times x + 4$ equal $4 \times (x + 1)$?* Students will probably agree that the expressions would still be equal, because each expression represents the number of tiles in the same region. Some might even do a few calculations with various values of x.

New York City Title I Elementary School Activities with The Geometer's Sketchpad
© 2012 Key Curriculum Press

5. *You can use Sketchpad to check for equivalence.* Go to page "Square." Mention that the displayed expression is an algebraic expression because it involves a variable. *Algebraic expressions are equivalent if they have the same values when the variable is replaced with any number.* Drag the x slider under the square to show that it controls the length of a side.

6. *Let's add another equivalent expression to the sketch. Which one shall we add from our list?* For example, if students pick $4 \times (10 + 2) - 4$, show how to choose **Number | Calculate** and enter into the Calculator $4 * (x + 2) - 4$. Then drag the slider for the square's side length to show that the values are the same. *What's the word again for expressions that have the same values for every number you substitute?* [Equivalent]

DEVELOP

Expect students at computers to work about 20 minutes.

7. Assign students to computers and tell them where to find **Pool Border.gsp**. Distribute the worksheet. Tell students to work through step 6 and do the Explore More if they have time.

8. As you circulate, help students enter new expressions as needed. Make sure they are drawing a diagram with shaded regions to illustrate each expression, but don't insist on any particular means of shading. Some of the diagrams may be ambiguous, but students should be able to explain their own drawings.

9. Have students put selected expressions and the accompanying diagrams on the board. Where diagrams don't represent the thinking clearly, ask for explanations. Students might display these expressions.

   - $4x + 4$
   - $4(x + 2) - 4$
   - $4(x + 1)$
   - $2(x + 2 + x)$
   - $2(2(x + 2) - 2)$
   - $(x - 2)^2 - x^2$

10. Ask students who worked on page "Rectangle" to share expressions for the number of border tiles in terms of side lengths m and n of the rectangle.

SUMMARIZE

Expect to spend about
10 minutes.

11. Reconvene the class. Have volunteers show their diagrams and explain their expressions.

12. ***Let's use algebra to change expressions into other equivalent expressions. How might you change*** $4(x + 1)$ ***into the expression*** $4x + 4$***?*** Students might say, *Multiply through.* Review or introduce the term *distribute.*

13. Now work with an expression in which students combine like terms. ***How might you change*** $2(x + 2) + 2x$ ***into the expression*** $4x + 4$***?*** Introduce or review the terminology *combining like terms.*

14. ***What have you learned?*** List all ideas students have. Bring out these objectives.

   • An algebraic expression involves variables.

   • If a number is written next to a variable, it means multiply.

   • An algebraic expression evaluates to a number when each variable is replaced by a number.

   • Algebraic expressions are equivalent if they have the same values when the variables are replaced by any number.

   • Distributing multiplication over addition gives equivalent algebraic expressions.

   • Combining like terms gives equivalent algebraic expressions.

15. ***What other questions can you ask that might be explored?*** Again, ideas will vary. Here are some interesting possibilities. You might suggest that students investigate these on their own.

   Why does distributing give equivalent expressions?

   Does addition distribute over multiplication?

   Do any other operations distribute over any operations?

EXTEND

If students are quite comfortable with distributing and combining like terms, you might ask **Can you distribute in the expression $2(2(x + 2) - 2)$?** You might have half the class take on that expression and half work with $(x + 2)^2 - x^2$. Or, you might try reversing the distributivity to introduce simple factoring. **How might you start with $4x - 4$ and get to $4(x - 1)$?** Note the common factor of both terms and using it as a multiplier.

ANSWERS

Answers to steps 2, 4, and 5–7 will include expressions and diagrams for $4x + 4$, $4(x + 2) - 4$, $4(x + 1)$, $2(x + 2) + 2x$, $(x + 2)^2 - x^2$, and other expressions. For reference, go to page "Expressions."

Answer to step 8 will include expressions and diagrams for $2m + 2n + 4$, $2(m + 2) + 2n$, $2m + 2(n + 2)$, $2(m + 1) + 2(n + 1)$, and other expressions.

Pool Border

For GSP5 Name:

In this activity you'll look for equivalent algebraic expressions for the number of tiles in a pool border.

EXPLORE

1. Open **Pool Border.gsp** and go to page "Square." One algebraic expression is already shown.

2. Look at the diagram of the pool and think of another way to combine the tiles. Choose **Number | Calculate** and enter an algebraic expression that represents that way of thinking. To enter x into the Calculator, click on the value in the sketch. What is your expression?

3. To change the size of the pool, drag the slider for x. Check that the value of your expression matches the value of the expression above. If not, change your expression until it does.

4. In the space below, draw a diagram that shows how you thought about the tiles. Write the expression under the diagram.

5. Repeat steps 2–4 for another expression. Draw a diagram and write the expression under the diagram.

New York City Title I Elementary School Activities with The Geometer's Sketchpad
© 2012 Key Curriculum Press

6. Repeat steps 2–4 for another expression. Draw a diagram and write the expression under the diagram.

7. Repeat steps 2–4 for another expression. Draw a diagram and write the expression under the diagram.

8. Go to page "Rectangle." Now you can change both the width and the length of the pool. One algebraic expression for the number of border tiles is given. Add as many other equivalent expressions as you can think of. Check that they give the same values as the given expression when you drag the sliders. List your expressions.

EXPLORE MORE

9. Go to page "Expressions." Figure out the expression for each diagram, then press *Show* to check your answer. Finally, use algebra to change the expressions into the other expressions.

Grade 5 Activities

How Close Can You Get: Rounding Decimals

INTRODUCE

Project the sketch for viewing by the class. Expect to spend about 5 minutes.

1. Open **How Close Can You Get.gsp.** Go to page "Rounding."

2. Explain, *Today you'll be looking at how different ways of rounding a number can produce dramatically different results. For example, rounding 3.499 to the nearest unit will produce 3, but so will rounding 2.501 to the nearest unit! In addition, depending on the number, sometimes rounding it to the nearest tenth will give the same value as rounding it to the nearest hundredth. But not always! Before you begin, I'll demonstrate how the sketch works.*

3. Drag the marker up and down the vertical axis. Ask students to pay attention to the values given by Hanna. Show students how to move the unit point or drag the tick mark labels so that they can zoom in or out on the number line. Stop dragging the marker and ask, *What is the value of the marker?* Many students will read off the value given by Hanna. If so, encourage students to explain how they can be sure that Hanna's value is the real value. Provide a counterexample if necessary (*So, if Hanna says 1.25, how can you be sure that the marker is not actually at 1.251?*). Make sure students understand that Hanna is reporting a value to the nearest hundredth.

4. *Now you'll investigate the other buttons shown on the sketch and compare the values they report as you drag the marker along the number line.*

DEVELOP

Expect students at computers to spend about 15 minutes.

5. Assign students to computers and tell them where to locate **How Close Can You Get.gsp.** Distribute the worksheet. Tell students to work through step 6 and do the Explore More if they have time.

6. Let pairs work at their own pace. As you circulate, here are some things to notice.

   • For worksheet step 3, help students make the argument that there are several different numbers Theresa could report that round to 1.25. Help students determine the interval of these values by identifying its maximum and minimum values.

   • For worksheet steps 5 and 6, students can drag the marker to help them solve the problems. Students may find it easier to move the marker using the arrow keys on their keyboards. They may also find it

SUMMARIZE

Project the sketch. Expect to spend about 10 minutes.

7. Bring students together. Ask students to talk about whether it is possible for Hanna to report 2.49 and for Ursula to report 3. Some students may think (incorrectly) that it is possible to "double" round so that 2.49 can be rounded to 2.5, which would then be rounded to 3.

8. Ask students to describe what happens to the four reported values as the marker is dragged from 2 to 3 in terms of their relative values. When is one bigger or smaller than the other? When are any pairs of values the same?

EXTEND

Ask students to describe different situations in which they use rounded values. [Rulers, calculators, weight on a scale, etc.] You might also ask them to think how rounding can be used in a misleading way.

ANSWERS

2. a. 1
   b. 1.2 or 1.3
   c. Answers will vary.

3. Theresa could report one of many values (from 1.246 to 1.254) that would be rounded to the nearest hundredth, as reported by Hanna.

4. Answers will vary.

5. a. Answers will vary. Sample solution: 0.951 and 1.049
   b. Answers will vary. Sample solution: 1.590, 0.400. Ursula is always either smaller or bigger than (never between) both Hanna and Theresa (unless the number is exactly an integer, in which case all three report the same value).
   c. Answers will vary. Sample solution: 0.403, 0.798

6. a. False
   b. True
   c. False
   d. True

7. a. No
   b. 1
   c. From 2.001 to 3.999
   d. When the units digit is odd

How Close Can You Get?

Name:

In this activity you'll explore how results differ when a number is rounded to the nearest unit, tenths, hundredths, or thousandths.

EXPLORE

1. Open **How Close Can You Get.gsp** and go to page "Rounding." You will see a vertical line with a marker on it. Drag the marker and observe how the value of Hanna changes.

2. Place the marker so that Hanna reports 1.25. Predict what the following people will report:
   a. Ursula, who reports to the nearest unit
   b. Tony, who reports to the nearest tenth
   c. Theresa, who reports to the nearest thousandth

3. Press *Show U, Show T,* and *Show Th* to check your answers for Ursula, Tony, and Theresa in step 2. You were probably wrong for Theresa. Why?

4. Press *Hide U, Hide H,* and *Hide Th,* and then press *New Location.* Write down the value that Tony reports. Then write down three different values that Hanna could report and three different values that Theresa could report.
   a. Tony
   b. Hanna
   c. Theresa

5. Press buttons to show all values. Drag the red marker along the vertical line.
   a. Find two locations where Ursula and Tony report the same numbers. Report your locations using Theresa's value.

   b. Find two locations where Hanna and Theresa report the same numbers. Report your locations using Theresa's value. What do you notice about Ursula's value? Explain.

   c. Find two locations where Tony and Hanna report the same numbers, but Ursula reports a different number. Report your locations using Theresa's value.

6. Drag the marker to help you decide whether the following statements are true or false.

   a. Theresa always reports a number greater than Ursula reports.

   b. Tony always reports a number that is either bigger than the numbers Hanna and Theresa both report or smaller than the numbers they both report.

   c. If Hanna reports 1.76, then it is possible that Theresa reports 1.750.

   d. The greatest possible difference between the numbers that Tony and Hanna report is 0.05.

EXPLORE MORE

7. Go to page "Explore More." Suppose that on another planet, people report their measurements to the nearest even number (Eva) or odd number (Odette) rather than to the nearest decimal.

   a. Can Eva and Odette ever report the same numbers?

   b. What is the biggest possible difference between the values that Eva and Odette report?

   c. If Odette reports 3, what is the range that the marker could be showing?

   d. When will Eva report a number greater than Odette?

Combination Locks:
Factors, Composites, and Primes

ACTIVITY NOTES

INTRODUCE

Project the sketch on a large-screen display for viewing by the class. Expect to spend about 20 minutes.

1. Begin by introducing the Sketchpad model. Open **Combination Locks Factors.gsp** and go to page "Bicycle Lock."

   Here is a very unusual bicycle lock. The lock has two dials. Right now, there are 12 tick marks along the red dial and 5 tick marks along the blue dial. Both dials have pointers that begin pointing up.

   When we set the pointers in motion, the pointer on the red dial will go all the way around its dial and end up back where it started. In this case, that means it will advance 12 times, jumping from one tick mark to the next.

   The pointer on the blue dial will advance the same number of tick marks as the pointer on the red dial. That's 12, in this case.

   Let's watch. Press *Go Slowly* and watch the pointers advance. The red pointer will make one full trip around its dial, while the blue pointer will make a little more than two trips around its dial. *Where are the pointers now?* [The red pointer points up when the movement stops, but the blue pointer does not.] *The lock did not open. For the lock to open, both pointers must point up when they stop.*

DEVELOP

Continue to project the sketch. Expect to spend about 20 minutes.

2. *Let's try to open the lock.* Press *Reset* to return both pointers to their starting positions. *We'll leave the red dial set to 12 tick marks. We'll change the number of tick marks on the blue dial. What number should we try?* Take suggestions. Double-click the blue *Ticks* value, enter a suggested number, and click OK. Press *Go Slowly* to set the pointers in motion. Again, they will advance 12 times.

Remember to press *Reset* before each new attempt to open the lock.

3. Try new values until the class finds one or more that open the lock. Suppose, for example, students discover that using two tick marks on the blue dial allows the lock to open. *How many times did the pointer on the blue dial go all the way around?* It's likely that students will need to watch the pointers move again to answer this question. Press *Reset* and then press *Go Slowly* again.

242

New York City Title I Elementary School Activities Grades 1–5
© 2012 Key Curriculum Press

4. On the board or chart paper, start a table with the column heads shown here, and add the data for this successful attempt to open the lock.

Red Tick Marks	Blue Tick Marks	Blue Trips Around the Dial
12	2	6

Find other numbers of blue tick marks that allow the lock to open. Record the data in the table. Here is a complete list.

One way to think about a dial with a single tick mark is this: The pointer goes once around the entire dial for each time it advances.

Red Tick Marks	Blue Tick Marks	Trips Around the Blue Dial
12	1	12
12	2	6
12	3	4
12	4	3
12	6	2
12	12	1

5. Discuss the data the class has recorded. ***Do you see a connection between the numbers in each row of the list?*** Through discussion, elicit the idea that 12, the number of red tick marks, is equal to the number of blue tick marks times the total trips around the blue dial ($12 = 1 \times 12$, $12 = 2 \times 6$, $12 = 3 \times 4$, $12 = 4 \times 3$, $12 = 6 \times 2$, $12 = 12 \times 1$).

Students may also note that the third column is the "reverse" of the second column. ***Why is that so?*** Here are sample student responses.

If you look at each row of numbers, the second and third columns have all the pairs of factors for 12. Multiplication is commutative, so there are two ways to get around the dial for each factor pair. When there's a 4 in the second column, there's a 3 in the third column. When there's a 3 in the second column, there's a 4 in the last column.

The number of trips around the blue dial is always equal to 12 divided by the number of blue tick marks. If I have four ticks, that's three trips. If I have three ticks, that's four trips. The numbers flip.

When students become more adept at reasoning with the model, you can press *Go Quickly* to speed the action.

6. Try new numbers of tick marks for the red dial. Double-click the red *Ticks* value, enter a number students suggest, and click OK. Again, have students suggest numbers of tick marks for the blue dial that will open the lock. Don't try any prime numbers for the red dial at this time; students will explore primes in step 7. Before testing a number for the

blue dial, ask, ***How do you know that this number will open the lock? Who agrees?*** Continue trying different locks until students are confident they can determine a number of blue tick marks that will open the lock for any number of tick marks on the red dial.

SUMMARIZE

Continue to project the sketch. Expect to spend about 20 minutes.

7. Double-click the *Ticks* value for the red dial and change it to 13. Ask students to consider what number of tick marks on the blue dial will allow the lock to open now. Give students an opportunity to suggest and try different numbers. [Because 13 has no factors other than 1 and itself, the blue dial must be set to either 1 or 13 ticks.]

   The number 13 was different from 12. We found quite a few numbers that opened the lock for 12, but only two for the number 13. Can you think of numbers other than 13 that would have this same property? Facilitate discussion, recording numbers the class agrees upon. [Any prime number—a number with exactly two different factors (itself and 1)—will share this same behavior.] You can use this opportunity to introduce the terms *prime number* and *composite number*. A composite number is one with factors other than 1 and itself.

8. ***How would you explain to someone else how to determine a number of tick marks for the blue dial that will open the lock? Why do some numbers work, whereas others do not?*** Have students talk in pairs, and then discuss as a class. Facilitate as students develop the explanation that the pointer on the red dial will always move exactly once around its dial and point up when it stops. For the blue dial to end pointing up, it needs to make some number of complete trips around, with no extra advances. Numbers that divide evenly into the number of tick marks on the red dial—factors of that number—result in complete trips, with nothing left over.

9. Discuss the following question. Alternatively, offer it as a writing prompt and have students respond individually. ***I want to make a lock that is very easy to open. It should open for lots of different numbers of tick marks on the blue dial. How many tick marks do you think I should use for the red dial? What would be a good choice?*** [Good choices are numbers like 16, 18, 20, 24, 28, and 30, all of which have at least five factors.]

EXTEND

1. For students who would benefit from individualized work with the sketch, provide an opportunity to use it alone or in pairs. Provide numbers to enter for the red dial, such as 15, 36, and 17. For each problem, have students determine all the different numbers of tick marks for the blue dial that will allow the lock to open. Alternatively, partners can create bicycle-lock challenges for each other by changing the number of tick marks on the red dial.

2. *What questions would you like to pose about this bicycle lock?* Encourage all student inquiry. Mathematical questions of interest include these.

   - Is there a pattern that allows us to predict how many factors a number has?

   - Which number less than 100 has the most factors?

   - Are all numbers either prime or composite?

   - Are fractions factors of anything?

   - Are negative numbers factors of anything?

   - What if one of the dials is set to 0 tick marks? [The sketch does not accommodate this.]

3. As a challenge, pose this question: *Suppose you could choose any number of tick marks between 1 and 200 for the red dial. Which do you think would make it most difficult for someone to open the lock?* Given these conditions, it's tempting to suggest a large number like 199 for the tick marks. But this lock will open easily by setting the number of the tick marks on the blue dial to 1 or 199.

   After students have recognized this possibility, add one more condition to the problem. *For higher security, your lock has a new feature: It will not open if the number of tick marks on the blue dial is exactly equal to the number of tick marks on the red dial or if it is equal to 1.* An immediate consequence of this condition is that if the number of tick marks on the red dial is prime, the lock won't open at all! This is too restrictive an option, as even the owner of the lock could not open it. Confine the search to composite numbers.

Students might still think that numbers very close to 200 are good choices. This is not necessarily the case. If the red dial is set to 198 tick marks, the lock opens when the blue dial is set to 2 tick marks. Ideally, the number of tick marks on the red dial needs to have factors that are hard to find. Possibilities include $11 \times 13 = 143$, $13 \times 13 = 169$, and $11 \times 17 = 189$.

How Many Bugs: Divisibility and Remainders

ACTIVITY NOTES

INTRODUCE

Project the sketch on a shared computer for viewing by the class. Expect to spend about 10 minutes.

1. Open **How Many Bugs.gsp.** Go to page "Group Bugs." Explain, *Each orange circle represents a bug. Notice it says there are 44 bugs. It's hard to count them, though, because they're scattered. Let's group the bugs together so they're easier to count.* Press *Group Bugs.* The bugs will assemble themselves into 7 groups of 6 bugs each. Two extra bugs will remain. Be sure students see the connection between the "Bugs Per Group = 6" parameter and the number of orange circles in each group. Ask students to describe how to find the total number of bugs now. [$7 \times 6 + 2$] Review that 2 is the *remainder* when 44 is divided by 6.

2. Press *Scatter Bugs.* Double-click *Bugs Per Group* and change the number to 8. Click OK. *Predict how the bugs will assemble themselves now that the number of bugs per group has changed.* Provide thinking time. Press *Group Bugs* and note that there are now 5 groups of 8 bugs with 4 bugs remaining. Have students complete the equation, $44 = \underline{} \times 8 + \underline{}$. *We can also say that 4 is the remainder when 44 is divided by 8.*

DEVELOP

Continue to project the sketch. Expect to spend about 30 minutes.

3. Go to page "How Many Bugs?" *Unlike the previous model, now the number of bugs is hidden. The goal is to figure out how many bugs appear on screen by collecting some clues.* (For your information, there are 53 bugs.) Explain that as before, the bugs will group themselves based on the number for Bugs Per Group. Ask students how many bugs they'd like per group, and change the number on screen to the chosen value. For our example, we'll keep the number of bugs per group at 6. (For spacing purposes, it's best to keep the number of bugs per group less than 14.)

   Press *Group Bugs.* The bugs will assemble themselves into groups, but immediately afterward, a black wedge will appear, covering the groups. All that can be seen are those bugs that belong to the remainder. (If there is no remainder, the entire circle will be black.) In our example, only 5 bugs are visible. (8 groups of 6 bugs are hidden.)

4. *I tried to count the groups of bugs, but they disappeared too quickly! All we can see now is how many bugs remain. When the bugs are divided into groups of 6, there are 5 bugs left over.*

Do you think this information can help us? Even though we don't know how many groups of bugs there are, does it help to know the number of bugs per group and the remainder?

Some students may claim they know how many groups of bugs are covered, but this is at best a guess.

Give students a few minutes to talk in pairs, and then discuss the question as a class. Students are likely to suggest specific numbers that leave a remainder of 5 when divided by 6. Record these. *Let's list other numbers that are 5 more than a multiple of 6.* Have the class list the first few numbers, in order (5, 11, 17, 23, 29). *Is there an easy way to find and list the numbers that leave a remainder of 5? Study this sequence. Can you find any relationships between the numbers?* [The numbers increase by 6—the number of bugs per group.] Have the class continue the list to include 35, 41, 47, 53, 59, 65, 71, 77, 83, and 89.

5. Press *Scatter Bugs* to disperse the bugs again. *Let's divide the bugs into a different number of bugs per group.* Take a suggestion and enter the number. For the sake of example here, we'll assume the number of bugs per group becomes 5. Press *Group Bugs*. In our example, now only 3 bugs are uncovered. (10 groups of 5 bugs each are hidden.)

   Discuss as a class what new information can be deduced. In our example, dividing the bugs into groups of 5 leaves a remainder of 3. That means the total number of bugs is 3 more than some multiple of 5. What are some numbers that are 3 more than a multiple of 5? Possibilities include 3, 8, 13, 18, 23, 28, 33, 38, 43, 48, 53, 58, 63, 68, 73, 78, 83, 88, and so on.

6. *Now that we have two lists of numbers, what do we know?* Give students a few minutes to talk in pairs, and then discuss the question as a class. Elicit the following ideas.

The lists will share other numbers too, if the class has extended them further.

   • Comparing the lists of possible bugs when grouped by 6 and grouped by 5 (as in our example) reveals three numbers in common: 23, 53, and 83. All three numbers when divided into groups of 6 leave a remainder of 5, and all three numbers when divided into groups of 5 leave a remainder of 3. (Remind students that the total number of bugs has stayed the same.)

   • It might be possible to take an educated guess about how many bugs are on screen by thinking about whether the number of scattered bugs appears closer to 23, 53, or 83.

• A more logical way to gather more information is to divide the bugs into another size of group and look at the remainder. ***Is there a group size that will eliminate two of the three choices?*** Distribute paper and let students explore this by carrying out some division by hand. In our example, the three possibilities have different remainders when divided by 8. In the sketch, dividing the bugs into groups of 8 leaves a remainder of 5. That information fits one of the possible totals, 53 $(6 \times 8 + 5)$, but not 23 or 83 bugs.

7. Have the class try one or more numbers to group by and work with the results to identify the total number of bugs.

> To keep the bugs well spaced on screen, it is best to restrict their number to no more than 100.

8. Go to page "Make Your Own." While a volunteer holds a hand in front of the projector lens, enter a new value for *Number of Bugs.* Alternatively, invite another volunteer to enter the new number. With the new number of bugs entered, press *Hide Number of Bugs.* The model can now be shown to the rest of the class and is ready for solving. As before, students repeat the following three steps, in order, until they have enough information to deduce how many bugs are on screen.

   • Change the number of bugs per group.

   • Group the bugs (viewing only the remainder).

   • Scatter the bugs.

SUMMARIZE

> Working away from the computer if you wish, expect to spend about 20 minutes.

9. Ask students to consider how they were able to solve the bug problems. ***You figured out how many bugs were on the screen even though nearly all the bugs were covered. What kinds of clues did you use to solve the problems? What were your strategies?*** Elicit the following key observations.

   • The clues were the number of bugs per group and the number of bugs remaining after they grouped themselves together. More concretely, if there were 53 bugs divided into groups of 6, then $53 \div 6 = 8$ remainder 5. But students knew only two pieces of information: the divisor (6) and the remainder (5).

- The strategies to solve the bug problems included working backward (using the divisor and the remainder to reconstruct what was being divided) and process of elimination (comparing lists to see what numbers they shared in common).

EXTEND

1. For students who would benefit from more individualized work with the bugs model, consider giving them an opportunity to use page "Make Your Own" alone or in pairs at a later time.

2. ***What questions about how-many-bugs problems would you like to pose? What are you wondering about?*** Encourage all curiosity. Here are some questions students might pose.

   If I know remainders for some group sizes, can I predict remainders for other group sizes?

   What's the least number of tests we need to make?

   Is there a systematic way of choosing group sizes that eliminates lots of possibilities?

   If we try the group size 3, is it useful to try group sizes that are multiples of 3, such as group sizes of 6 or of 9? Will we learn anything new?

Mystery Number: Multiples and Factors

For GSP5 ACTIVITY NOTES

INTRODUCE

Project the sketch on a large-screen display for viewing by the class. Expect to spend about 15 minutes.

1. Open **Mystery Number.gsp.** Go to page "Mystery Number." Distribute the worksheet. Explain that the class's challenge is to find the computer's mystery number. *The mystery number is a number from 1 through 25. We can ask for clues.* Press the *Multiple of 2?* button. A check mark appears in the Yes column. *What did we learn by pressing this button?* [The mystery number is a multiple of 2.]

   In step 1 of the worksheet, have students enter the button pressed (2) and the clue (Yes) in the chart.

2. Explain that solving the puzzle requires careful reasoning. *Our goal is to figure out the mystery number using as few clues as possible. Let's think about the information we've been given and see whether it helps us narrow down our choices. What can you say about the mystery number now that you know it is a multiple of 2?* Here are some sample student responses.

   It must be an even number.

   It can't be an odd number.

   It could be 2, 4, 6, 8, 10, 12, 14, 16, 18, 20, 22, or 24.

   On your worksheets, I'd like you to use the list under the chart. Cross off the numbers that cannot be the mystery number. Students should point out that these are all the odd numbers in the list.

Some students may not think that 2 is a multiple of 2. Since $2 \times 1 = 2$, however, 2 is a multiple of itself.

3. Continue working on the puzzle. *I'm going to press another button.* Press *Multiple of 9?* A check mark appears under Yes. Ask students to enter this new information on their worksheets. *What can you say about the mystery number now?* Have students talk with a partner, and then take responses. The mystery number is a multiple of both 2 and 9, so it must be 18. When the class is convinced that the number is 18, demonstrate pressing *Show mystery number* to check.

As you discuss students' ideas, make a point of using their own language. Also, incorporate terms such as *even, odd, and multiple* if students don't use them.

4. Draw students' attention to the *Show Sum of the Digits* button. Press the button. It displays the sum of the digits in the mystery number—in this case, $1 + 8 = 9$. Explain that this button should be pressed for a hint as a last resort. If students have pressed every button in the table and they are not able to determine the mystery number, they should press this button.

DEVELOP

Continue to project the sketch. Expect to spend about 30 minutes.

5. Solve another puzzle together. Press *New Puzzle*. Explain that the computer generates a new mystery number from 1 to 25 at random. Ask a volunteer to come to the computer and choose a button to press. Have students record the button and Yes/No clue in the chart in worksheet step 2.

6. Facilitate a discussion of the clue by asking questions such those below. Use the list of the numbers 1–25 on the board and have students use the list on the worksheet to keep track of numbers that have been eliminated and numbers that remain.

   Are there any numbers that cannot be the mystery number? How do you know?

   What numbers might be the mystery number? How do you know?

   Now that you have this clue, do you know the answers to any of the other questions (buttons)?

7. Ask volunteers to press a button for another clue and for any other clues the class may need in order to find the mystery number. As students think about the information they are obtaining, many different mathematical conversations can develop. Here are some scenarios that suggest the reasoning students may use.

   • Student presses *Multiple of 8?* The answer is No.

   Since the mystery number is not a multiple of 8, it can't be 16 or 24. But it's possible that the mystery number is a multiple of 2 or 4 (both are factors of 8).

   • Student presses *Multiple of 10?* The answer is Yes.

   Every number that is a multiple of 10 is also a multiple of 2 and of 5. The mystery number must be a multiple of both 2 and 5. So we don't need to check those buttons. Since the number is a multiple of 10, it could be either 10 or 20. To check whether the number is 20, we could press Multiple of 4.

- Student presses *Multiple of 7?* The answer is Yes.

  The possibilities are 7, 14, and 21. We can check to see whether the number is a multiple of 2 or a multiple of 3. If it's a multiple of 3, it has to be 21. If it's a multiple of 2, it has to be 14. If it's not a multiple of 2 or of 3, it has to be 7.

- All buttons are pressed and the answers are all No.

  The numbers that have buttons (2 through 12) can be crossed out. The numbers 14, 15, 16, 18, 20, 21, 22, 24, and 25 are all multiples of numbers that are less than or equal to 12, so they can be crossed out. That leaves the prime numbers 13, 17, 19, and 23 as well as 1. The only way to tell which of these numbers is the mystery number is to press Show Sum of the Digits.

- The mystery number is a multiple of 5, but no other number.

  The number could either be 5 or 25. There is no way to tell which it is without pressing Show Sum of the Digits.

8. If time allows, solve more puzzles. Each puzzle will expose students to new and interesting properties of multiples and factors.

SUMMARIZE

Continue to project the sketch. Expect to spend about 15 minutes.

9. Facilitate discussion of the strategies students used to find the mystery numbers. Students can choose examples from their worksheet and describe, step-by-step, the reasoning that helped them to deduce the mystery number.

10. *To find the mystery numbers, you had to do a lot of thinking about multiples that numbers have in common. I have two questions for you.* Facilitate as the class discusses each question.

    Why is a multiple of 10 also a multiple of 2 and 5?

    If you know that a number is not a multiple of 10, do you know that it is not a multiple of 2 and 5?

11. Pose one or more of the problems that follow and have students work in pairs and then share solutions with the class. Alternatively, have students write individually in response to one or more problems. Students will notice that now the mystery number can be a number as large as 30.

- Ann is thinking of a number between 1 and 30. She says the number is a multiple of 8 and a multiple of 3. What is the number? Explain. [24, the only multiple of 8 and 3 that is between 1 and 30]

- Hector is thinking of a number between 1 and 30. He says it is a multiple of 12. You want to know whether the number he is thinking of is 24. What one question about multiples can you ask him to find out? Explain your thinking. [Is the number a multiple of 8? If it is, then the number is 24 and not 12.]

- Sara is thinking of a number between 1 and 30. It has no factors between 2 and 15. What number could she be thinking of? Is there only one possibility? [Sara could be thinking of any prime number between 15 and 30. These are 17, 19, 23, and 29. She could also be thinking of the number 1.]

- Daniel is thinking of a number between 1 and 30. He tells you that it has more factors than any other number in the puzzle. What number is she thinking of? Explain how you know. [The number 24 has more factors than any other number between 1 and 30. Its factors are 1, 2, 3, 4, 6, 8, 12, and 24.]

EXTEND

1. Provide an opportunity for students to solve puzzles in pairs or individually. Students can create puzzles for others to solve using page "Make Your Own." If some students would like to play for points, suggest that they give themselves 1 point for each clue they use to find the mystery number. A low score for solving the puzzles is a good score.

2. *What questions would you like to pose about mystery-number puzzles?* Encourage student inquiry. Some mathematical questions of interest include the ones here.

   - What would happen if we showed both the mystery number and all the checkmarks and then dragged the slider? What would we see?

   - What numbers in the 1–25 range require the fewest clues to find? Which numbers require the most clues?

- What if the computer chose numbers up to 50? Would we need questions about other multiples to be able to identify each number?

- Are there other kinds of questions that might allow us to figure out the numbers with fewer clues?

ANSWERS

1–4. Answers will vary because the computer randomly generates mystery numbers and because students' solutions will vary.

Mystery Number

For GSP5

Name: _____

Solve each puzzle using as few clues as you can.

EXPLORE

1. Mystery number = _____
   Record the buttons you press and the clues.

Multiple of	Yes	No

Multiple of	Yes	No

Cross off numbers that can't be the mystery number.
Circle numbers that might be.

1   3   5   7   9   11   13   15   17   19   21   23   25

2   4   6   8   10   12   14   16   18   20   22   24

2. Mystery number = _____

Multiple of	Yes	No

Multiple of	Yes	No

1   3   5   7   9   11   13   15   17   19   21   23   25

2   4   6   8   10   12   14   16   18   20   22   24

New York City Title I Elementary School Activities Grades 1–5
© 2012 Key Curriculum Press

3. Mystery number = _____

Multiple of	Yes	No

Multiple of	Yes	No

1   3   5   7   9   11   13   15   17   19   21   23   25

2   4   6   8   10   12   14   16   18   20   22   24

4. Mystery number = _____

Multiple of	Yes	No

Multiple of	Yes	No

1   3   5   7   9   11   13   15   17   19   21   23   25

2   4   6   8   10   12   14   16   18   20   22   24

Two-Digit Multipliers: Visualizing the Distributive Property

INTRODUCE

Gather the class. Expect to spend about 20 minutes.

1. Present the following problem, which includes a faulty solution.

   The other day I needed to buy a lot of cookies for an event. At the store I saw that the cookies come in packages of 18. There were 15 packages on the shelf.

   As I was putting all the packages in my shopping cart, I wondered aloud whether this many packages would be enough. My niece, who was with me, said, "I can figure out how many there are: Multiply 10 times 10, and 5 times 8, and add them up."

   Write the following on the board as you relate the niece's strategy.

   $$15 \times 18 \qquad\qquad 10 \times 10 \qquad\qquad 5 \times 8$$

   What do you think? Do you agree with my niece? Have scratch paper available and provide time for all students to consider this strategy.

2. Facilitate a discussion in which students consider each other's thinking and reveal the erroneous thinking in the niece's solution.

Three strategies students might offer are multiply 18 by 30 and halve the result; multiply 15 by 20 and subtract two 15's; and multiply 18 by 10 and add half the result.

Some students may use mental computation strategies and find that the product of 15×18 is larger than the result of the niece's method, as in this example.

Fifteen times 20 is 300. That is 30 more than 15 times 18, but it's obvious that the result is going to be a lot more than 140, which is what your niece would get.

Other students are likely to explain that the niece's strategy is incomplete, as in these two examples.

You have to do more than just multiply 5 times 8. It's 5 times 18. So, you also have to multiply 5 times 10.

What if you bought 10 packages at one store and then you bought 5 more packages at another store? That would be the same as 15 packages. Ten packages would be 10 times 18, which is 10 times 10, plus 10 times 8. Five more packages would be 5 times 18. That would be 5 times 10, and 5 times 8.

3. The second response above presents the computation in a way that many students will be able to understand. If it is not proposed by students, introduce it. **Suppose I bought 10 packages at one store and 5 at another store. Is that the same as buying 15 packages at one store?**

How many cookies would I buy at each store? Make a recording like the one below as students share their thinking. Save the recording for use in the next step of the activity.

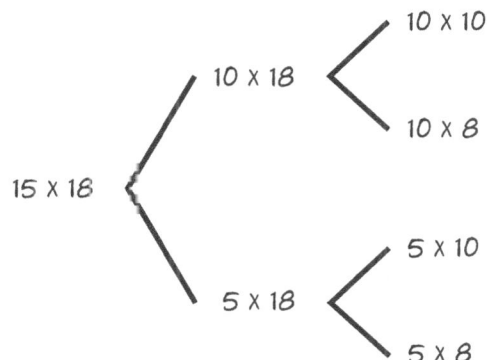

DEVELOP

Project the sketch for viewing by the class. Expect to spend about 20 minutes.

4. Open **Two-Digit Multipliers.gsp.** Go to page "Area Model." The problem students have been discussing, 15×18, should be displayed.

5. *How does this model show the strategies you've been talking about?* Students may point out that there are four small areas in the model and that these areas represent the four computations in the recording (as in step 3 above). Help all students make sense of the model. You may press *Show Grid* to help students connect the model to base ten blocks. Also, ask these questions.

   What does this large square represent? $[10 \times 10]$

   What does this set of rectangles represent? $[10 \times 8]$

   What about the other set of rectangles? $[5 \times 10]$

   And the small squares? $[5 \times 8]$

6. *How would you use this model to explain to my niece why her way of finding the product doesn't work?* Facilitate discussion, encouraging students to consider each other's explanations. Invite students to the computer to point to parts of the model as they share their explanations with the class.

7. Introduce a new problem. In the model, drag the point at the bottom-right corner and the point at the top-left corner to change the problem to 13×17.

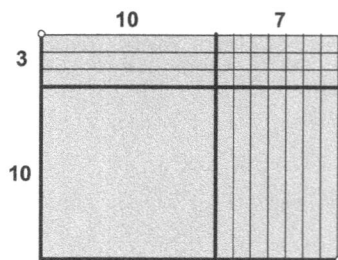

Give the class a moment to study the model. ***What would you write down to figure out how much 13 times 17 is?*** Ask volunteers to share their thinking on the board. Students will have varied ways of recording the partial products. Make sure students understand that any order of recording the partial products is correct. (For your information, the recording of the largest product first—*tens* times *tens*—is just as efficient as the standard expanded form, which begins by multiplying *ones* by *ones*.) Here is one strategy.

$$10 \times 10 = 100$$

$$10 \times 7 = 70$$

$$3 \times 10 = 30$$

$$3 \times 7 = 21$$

$$100 + 70 + 30 + 21 = 221$$

Explain that the products for the small regions are called the *partial products*, and the total of all the partial products is called the *product*. Make sure students are able to think about the word *partial* in a way that makes sense to them: Each partial product shows a *part* of the whole.

> Students frequently start with the largest numbers, the tens, and work down to the smallest numbers, the ones. Notice how this way of working is like a common mental computation strategy.

8. Distribute the worksheet. Explain that students should represent the Sketchpad model by making a "shorthand" diagram in step 1. Draw the diagram shown here. Have students copy it and write the partial products and product on the lines provided on the worksheet.

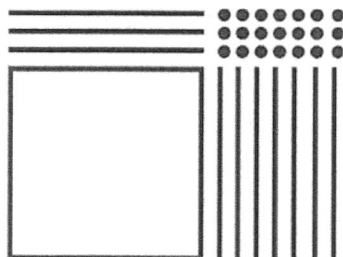

9. In the sketch, change the model to show the second worksheet problem, 16×23. Invite students to identify what is different: There are *two* hundreds squares. **Where did they come from?**

    One explanation students may offer is, *Twenty is two tens. Two tens times one ten is 10 times 10 twice. That's two hundreds.* Highlight the language, *two tens times one ten makes two hundreds.* It will help students describe the model and make sense of the computation.

    Have students make a shorthand drawing and record the partial products and product for 16×23.

10. Use the model to show the last problem on the worksheet, 17×33. Have students record in the same way.

SUMMARIZE

Continue to project the sketch. Expect to spend about 20 minutes.

11. Explain that the method students are using for carrying out these multiplication computations is called an *algorithm*. An algorithm is a step-by-step procedure that can be used for problems of a certain type, no matter what the numbers are.

    Make the recording shown at right, thinking aloud as you figure out each partial product.

    ```
           12
       X   45
       ------
          400
           80
           50
       +   10
       ------
          540
    ```

12. *You've learned an algorithm that will always work for finding the product of a two-digit number multiplied by a two-digit number. Does that mean you should always use this algorithm? When is a good time to use it?* Depending on how much experience students have had with mental computation, they may suggest that the method is good to use when an easier and faster mental computation strategy does not come to mind.

    As an example of a case where an alternative strategy might be faster, ask, **What strategy would you use to find the product for 15 multiplied by 19?** Students may suggest using 20, a "friendlier" number than 19. *I thought about* 20×15. *Ten times 15 is 150. So, 20 times 15 is 300. Then I took away one 15 because it was really 19, not 20, times 15. Three hundred minus 15 is 285.*

    When an efficient mental computation strategy such as the one above has been shared, ask, **Is that a fast way? Which would be faster—that way or the paper-and-pencil algorithm?** Students should note that the mental computation is probably faster. *The algorithm you've learned is a handy strategy to know, but it always makes sense to look at a problem and ask, "What's an easy way to do this?"*

13. With the class, make a poster about how and why the algorithm they have learned works and when to use it. The class might print a page of the area model and also show examples of problems for which the algorithm is and isn't an efficient strategy.

EXTEND

1. Some students may be ready to combine pairs of partial products mentally, reducing to *two* the number of partial products they record. Take students through the example shown at right. Then present other problems with numbers students are able to add mentally.

```
   12
 X 45
 480  (40 X 10) + (40 X 2)
+ 60  (5 X 10) + (5 X 2)
 540
```

2. Some students may benefit from more time with the sketch. Provide a set of two-digit multiplication problems and invite them to solve the problems in any way they want: They may use the sketch, a shorthand drawing, the algorithm the class explored, or mental computation strategies.

Two-Digit Multipliers

Name:

A. 13×17

B. 16×23

C. 17×33

Magic Multiplying Machine:
Exploring Multiplication

ACTIVITY NOTES

INTRODUCE

Project the sketch for viewing by the class. Expect to spend about 5 minutes.

1. Open **Magic Multiplying Machine.gsp.** Go to page "Multiplication Machine." Enlarge the document window so it fills most of the screen.

2. Explain, ***Today you're going to use Sketchpad to explore multiplication using a magic multiplication machine. What do you see?*** The sketch shows the multiplication problem $2 \cdot 5 = 10$. Students should observe that $a = 2$, $b = 5$, and $a \cdot b = 10$. The marker for a points to 2, the marker for b points to 5, and the marker for $a \cdot b$ points to 10 on the number line. The dot between a and b represents multiplication.

3. ***What do you think will happen if I drag marker a to the right?*** Have volunteers state their predictions. Then drag marker a to the right. The value of b stays the same. The value of $a \cdot b$ increases; it goes to the right.

4. ***What do you think will happen if I drag marker a to the left?*** Again have students share their predictions. Then drag marker a to the left. The value of b stays the same. The value of $a \cdot b$ decreases; it goes to the left.

5. ***You will use this multiplication machine to explore several multiplication situations. Drag the a and b markers to change their values. You can select a marker and use the right and left arrow keys on your keyboard to move it one pixel at a time.*** Model how to select a marker and move it using the arrow keys. The values change by five hundredths of a unit. ***As you use the sketch, think about what it is about multiplication that makes the machine behave like it does.***

DEVELOP

Expect students at computers to spend about 30 minutes.

6. Assign students to computers and tell them where to locate **Magic Multiplying Machine.gsp.** Distribute the worksheet. Tell students to work through step 23 and do the Explore More if they have time. Encourage students to ask their neighbors for help if they are having difficulty with Sketchpad.

7. Let pairs work at their own pace. As you circulate, here are some things to notice.

   • As students work, encourage them to answer the questions with detailed descriptions that include both observations and explanations.

- In worksheet step 4, the product is to the right of its factors. ***Is this true for any two factors?*** Have students predict before proceeding. They will investigate other instances in which the product is not always to the right of its factors. Students will learn that multiplying one number by another does not always result in a product that is greater than either factor.

- In worksheet step 7, encourage students to think of different types of answers to this question. For example, both factors might be positive integers, or one factor could be a positive integer and the other could be a decimal between 0 and 1.

- In worksheet step 8, if students find only one answer, prod them to find another location. ***Are you sure there isn't another location that works? Try other possibilities.***

- In worksheet step 9, if students need help, have them substitute a value for b and then figure out what a must be to make $-b = a \cdot b$. ***Move the marker for b to a negative number. Then can you move marker a so that a · b is the opposite of marker b?***

- In worksheet step 11, students may find it helpful to press *Zoom In*.

- In worksheet step 12, students explore the Multiplication Property of Zero, and in worksheet step 13, they explore the Multiplication Property of One (Identity Property of One). The number 1 is the *multiplicative identity*. Students may not know each by its formal name; that's okay at this time.

- In worksheet steps 17–23, students explore multiplication with different signs. Encourage students to observe how the multiplication machine behaves and to write in their own words why it reacts that way.

- If students have time for the Explore More, they will explore a mystery machine, a multiplication machine with the numbers left off the number line. Students are to find the locations of 0 and 1 by dragging a and b and observing the effect on $a \cdot b$. Have students verbalize their thinking as they try to solve the problem. They may press *New Problem* to try the challenge again.

SUMMARIZE

Project the sketch. Expect to spend about 10 minutes.

8. Gather the class. Students should have their worksheets with them. Begin the discussion by opening **Magic Multiplying Machine.gsp.** Review the worksheet questions with students. Have volunteers come up to the computer and explain their answers.

9. Summarize by asking students what they learned about multiplication by using the magic multiplication machine. ***How did the magic multiplication machine help you understand multiplication better?*** Here are some sample student responses.

   I always thought multiplying one number by another resulted in a larger number. Both factors need to be greater than 1 (or less than −1) for that to happen.

   It was clear that multiplying by 0 is always 0. When marker a was on 0, it didn't matter where I moved b. The marker for a · b stayed on 0.

   I thought it was pretty obvious that any number multiplied by 1 equals that number. I moved marker a to 1, and when I moved b, a · b always moved with it.

   The machine helped me understand why a negative times a positive is always a negative number. As I slowly dragged marker a to the left, I could see marker a · b following it.

10. If time permits, discuss the Explore More. Let volunteers share their strategies for finding the location of 0 and 1.

11. ***When you multiply two numbers, a and b, their product is positive. What can you say about the signs of a and b?*** You may wish to have students respond individually in writing to this prompt. The signs of the two numbers are the same: they are either both positive or they are both negative.

EXTEND

1. ***When you multiply two numbers, does the order matter? In other words, is −3 · 5 the same as 5 · (−3)? Using the sketch, explain why your answer makes sense.*** Students will learn that multiplication is commutative; the order of the factors is not important.

2. Show the table below with the right column blank. ***Determine what sign the product will have for each expression in the table. Assume that a, b, and c are positive. Can you make a conjecture about when a product is negative or positive?*** Students will discover that a product is positive when there are an even number of negative factors and negative when there are an odd number of negative factors.

Expression	Product Negative or Positive?
$a \cdot b \cdot c$	positive
$-a \cdot b \cdot c$	negative
$a \cdot (-b) \cdot c$	negative
$a \cdot b \cdot (-c)$	negative
$-a \cdot (-b) \cdot c$	positive
$a \cdot (-b) \cdot (-c)$	positive
$-a \cdot b \cdot (-c)$	positive
$-a \cdot (-b) \cdot (-c)$	negative

ANSWERS

4. The product $a \cdot b$ is to the right of the factors, a and b. This makes sense because the product of two numbers greater than 1 is always greater than either factor.

6. $0.5 \cdot 6 = 3$

7. Answers will vary. Sample answers include the following: $a = 2$ and $b = 3$, $a = -1$ and $b = -6$, $a = 1.5$ and $b = 4$, $a = -0.5$ and $b = -12$.

8. The only two possible locations are at 0 ($0 \cdot 0 = 0$) and 1 ($1 \cdot 1 = 1$).

9. a must be at -1.

10. No, not all three markers can be to the left of 0. When a and b are to the left of 0, $a \cdot b$ is to its right because a negative multiplied by a negative equals a positive. When $a \cdot b$ is to the left of 0, one of the factors is also to the left of 0 and the other is to the right because a negative must be the product of a negative and a positive.

11. Answers will vary. In all cases, both a and b are between 0 and 1.

12. Both a and $a \cdot b$ remain fixed at 0. This makes sense because any number multiplied by 0 equals 0 (Multiplication Property of Zero).

13. Markers b and $a \cdot b$ stay right on top of each other. This makes sense because any number multiplied by 1 equals itself (Multiplication Property of One).

14. Markers b and $a \cdot b$ move in opposite directions and at the same speed. They can be thought of as reflections of each other across 0. This makes sense because any number multiplied by -1 equals its opposite.

15. Marker b moves faster than $a \cdot b$—twice as fast, to be exact. This is because a number multiplied by a number between 0 and 1 gives a result closer to 0 than the original number.

16. The two possible answers are 2 and -2.

20. When b is positive, dragging a to the left also moves $a \cdot b$ to the left. As a approaches 0, $a \cdot b$ approaches 0, too, and when $a = 0$, $a \cdot b = 0$. It makes sense that as you keep dragging a to the left, $a \cdot b$ continues its previous behavior, moving to the left into negative territory.

23. With b to the left of 0, a and $a \cdot b$ move in opposite directions. As you drag a to the left, $a \cdot b$ moves to the right. As a approaches 0, $a \cdot b$ approaches 0, too, but from the other side, and when $a = 0$, $a \cdot b = 0$. It makes sense that as you keep dragging a to the left, $a \cdot b$ continues its previous behavior, moving to the right into positive territory.

25. One way to find 0 is to drag a (or b) until it and $a \cdot b$ are on top of each other. This spot is 0. This method works because any number multiplied by 0 equals 0. One way to find 1 is to drag a until b and $a \cdot b$ are on top of each other (or drag b until a and $a \cdot b$ are on top of each other). Marker a is at 1. This method works because any number multiplied by 1 equals itself.

Magic Multiplying Machine

For GSP5 Name:

In this activity you'll use a magic multiplication machine to explore the behavior of multiplication.

EXPLORE

1. Open **Magic Multiplying Machine.gsp** and go to page "Multiplication Machine."

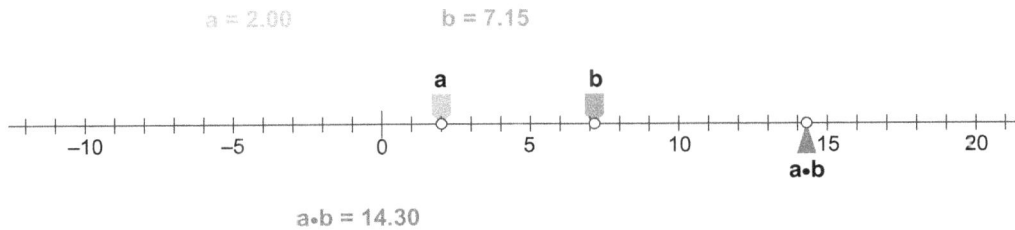

$a = 2.00$ $b = 7.15$

$a \cdot b = 14.30$

2. Drag points a and b. Observe what happens to $a \cdot b$.

3. Now drag points a and b to represent the problem $3 \cdot 4 = 12$.

   Use the right and left arrow keys on your keyboard to move the selected point one pixel at a time.

4. Where is the product compared to a and b? Explain why this makes sense.

5. Now drag a to 0.5 and b to 6.

6. What does the machine show for the product $0.5 \cdot 6$?

7. List four pairs of locations for a and b such that $a \cdot b = 6$.

8. Find a location for a and b in which a, b, and $a \cdot b$ all lie directly on top of each other. Is there more than one location that works?

9. If *b* and *a* · *b* are the same distance away from 0 but on opposite sides, where must *a* be?

10. Can *a, b,* and *a* · *b* all lie to the left of 0? Explain.

11. Find locations for *a* and *b* to the right of 0 such that *a* · *b* is less than both *a* and *b*. Describe all locations for *a* and *b* for which this works.

As you drag *a* or *b* to a value, observe the product *a* · *b* as it moves at the same time.

12. Drag *a* to 0. Then slowly drag *b* back and forth along the number line. What happens to the product *a* · *b*? Explain why this makes sense.

13. Drag *a* to 1. Then slowly drag *b* back and forth along the number line. Describe the movement of *a* · *b* in relation to *b*. Explain why this makes sense.

14. Drag *a* to −1. Then slowly drag *b* back and forth along the number line. Describe the movement of *a* · *b* in relation to *b*. Explain why this makes sense.

15. Drag *a* to 0.5. Then slowly drag *b* back and forth along the number line. Which moves faster, *b* or *a* · *b*? Explain why.

Magic Multiplying Machine
continued

16. Find a location for point *a* such that the distance from *a* · *b* to 0 is always twice the distance from *b* to 0. Are there other answers?

Have you ever wondered why a negative number multiplied by a negative number is a positive number? The magic multiplication machine provides a nice way to see the reason.

17. Drag *a* and *b* so that they're near the right edge of the sketch window.

18. Now drag *a* slowly to the left. Watch *a* · *b* move across the screen.

19. When *a* · *b* reaches the left edge of the sketch, move *a* in the opposite direction. Watch *a* · *b* move back to the right. Drag *a* back and forth and observe *a* · *b*.

20. Explain why it makes sense that a positive number multiplied by a negative number equals a negative number.

21. Next, drag *b* to the left of 0.

22. Once again, drag *a* back and forth. Watch *a* · *b*.

23. Explain why it makes sense that a negative number multiplied by a negative number equals a positive number.

Magic Multiplying Machine

continued

EXPLORE MORE

24. Go to page "Mystery Machine." This is a multiplication machine, but the numbers are hidden. Drag *a* and *b,* and try to find the location of 0 and 1 on the unmarked number line. Drag the gold arrows for 0 and 1 to mark your answer. Then press *Show Answers* to check. Press *New Problem* to try the challenge several times.

25. Describe the strategies you used to find the locations of 0 and 1.

Magic Dividing Machine:
Exploring Division

For GSP5

ACTIVITY NOTES

INTRODUCE

Project the sketch for viewing by the class. Expect to spend about 10 minutes.

1. Open **Magic Dividing Machine.gsp.** Go to page "Inverse Operations." Enlarge the document window so it fills most of the screen.

2. Explain, *Today you're going to use Sketchpad to explore division. You'll start by exploring this multiplication and division machine.* Slowly drag markers *a* and *b* to the right and left. *What's going on? Can you see how this model works?* Give students some time to form conjectures.

3. Drag marker *a*. *What happens when I drag marker a?* Students should see that marker *a* controls how many rectangles there are in the blue bar.

4. Drag marker *b*. *What happens when I drag marker b?* Students should observe that the width of the rectangles in the blue bar changes.

5. Press *Show Numbers*. Drag *a* to 2 and *b* to 4. *How many rectangles are there in the blue bar?* [2] *How wide is a rectangle in the blue bar?* [4 units] *Where is c?* [8] *Why does c have that value?* [2 bars end to end, each with length 4, have total length 8] *What multiplication problem does this represent?* [2 × 4 = 8]

6. *You can use the right and left arrow keys on your keyboard to move a marker one pixel at a time.* Model how to select a marker and move it using the arrow keys. The values change in five hundredths of a unit. *You can also use the two buttons in the lower-right corner to move a or b to the nearest integer.*

7. Model a few more multiplication problems, making sure students understand how the machine works. *Which marker represents a factor?* [*a* and *b*] *Which marker represents the product?* [*c*]

8. *How are multiplication and division related?* [Inverse operations] Press *Inverse Operation*. *Now the machine is in division mode. How is this different from the multiplication mode?* Students should see that marker *c* can now be dragged instead of marker *b*.

9. Press *Hide Numbers*. Drag *a* so that exactly two yellow squares are showing. Then slowly drag *a* so that exactly three yellow squares are showing. *What happened to the blue bar as I moved marker a from 2 squares to 3 squares?* [The blue bar went from being divided into two parts to being divided into three parts.] *You'll explore this model and another one much more on your own.*

DEVELOP

Expect students at computers to spend about 25 minutes.

10. Assign students to computers and tell them where to locate **Magic Dividing Machine.gsp.** Distribute the worksheet. Tell students to work through step 20 and do the Explore More if they have time. Encourage students to ask their neighbors for help if they are having difficulty with Sketchpad.

11. Let pairs work at their own pace. As you circulate, here are some things to notice.

   • In worksheet steps 1–9, students will work in multiplication mode. Check to be sure that students understand how it works before moving on to the division machine. **What does marker a represent? What does marker b represent? What is c?**

   • In worksheet step 10, students switch to the division mode. This is an important step. Encourage students to observe closely and to think about what happens during the switch. Ask students to press *Inverse Operation* several times so that they get a clear understanding of what changes. Students should notice that they can now control marker c but not marker b. Geometrically, students are now controlling the size of the blue rectangles by dragging the end of the bar (representing the product/dividend) rather than by dragging the end of the first rectangle (the multiplier/quotient).

   • In worksheet step 12, remind students that they can use the left and right arrow keys on their computers to move in small increments.

   • In worksheet step 14, students rewrite the multiplication equation as a division equation to calculate b. Product c now becomes the dividend; factor a becomes the divisor; and the other factor, or multiplier b, becomes the quotient: $c \div a = b$. Be sure students grasp this concept. **What does c represent now? What does a represent? How about b?** As needed, review the terms used in division: dividend, divisor, and quotient.

   • In worksheet step 15, students will use the machine in division mode to solve several problems. This will test students' understanding of how the model works and the relationship between multiplication and division.

- In worksheet steps 16 and 17, students work with another division machine on a number line, so that they can focus more on the behavior of division.

- In worksheet steps 18–20, students must translate the questions, which are in mathematical terms, into the terms of the model and the behavior of the markers; they must manipulate the model appropriately; and they must interpret the behavior they observe and express their conclusions mathematically. These questions help students develop their translation and interpretation skills.

- If students have time for the Explore More, they will explore a mystery machine, a division machine with the numbers left off the number line. Students are to find the locations of 0 and 1 by dragging a and c and observing the effect on $c \div a$. Have students verbalize their thinking as they try to solve the problem. They may press *New Problem* to try the challenge again.

SUMMARIZE

Project the sketch. Expect to spend about 10 minutes.

12. Gather the class. Students should have their worksheets with them. Begin the discussion by opening **Magic Dividing Machine.gsp** and going to page "Inverse Operations." ***What happens when you switch from the multiplication machine to the division machine? How are the two machines related?*** Here are some sample responses.

Marker a still controls the number of rectangles in the blue bar in both machines. Marker b is the width of a rectangle in the blue bar in both machines, but you can't move it in the division machine. Marker c is the length of the blue bar in both machines, but you can't move it in the multiplication machine. So the machines are the same except for what markers you can control.

In the multiplication machine, you multiply two factors, a and b, to find the product c. In the division machine, you can control the dividend and the divisor, c and a, to find the quotient b. This shows how multiplication and division are related: The product is the same as the dividend, a factor is like the divisor, and the other factor is like the quotient.

In the multiplication machine, you are finding the total length of the blue bar by multiplying the number of rectangles by the width of one rectangle.

In the division machine, you are finding the width of one rectangle by dividing the total length of the blue bar by the number of rectangles.

13. Work through the answers to problems 18–20 with the class. Have volunteers come up to the computer and explain their answers using the division machine.

14. ***How did the division machine help you understand division better?*** Here are some sample student responses.

    I always thought dividing by a positive number resulted in a smaller quotient. I could see that if you divide a positive number by a number between zero and one, you get a larger number.

    The machine helped me understand why a positive divided by a negative is always a negative number. As I slowly dragged marker a to the left, I could see marker c ÷ a jump to the left side as soon as marker a passed zero.

15. If time permits, discuss the Explore More. Have students explain what strategies they used to find the location of 0 and 1.

16. You may wish to have students respond individually in writing to this prompt. ***How can you find a missing factor in a multiplication equation? For example, in $5.12 \cdot b = 25.6$, how can you find b?*** Students should state that they can use division to find b: $25.6 \div 5.12 = b$, or $b = 5$.

EXTEND

1. ***When you divide two numbers, does the order matter? In other words, is $8 \div 2$ the same as $2 \div 8$?*** Using the sketch, explain why your answer makes sense. Students will learn that division is not commutative; the order in division is important.

2. ***What other questions might you ask about division?*** Encourage all inquiry. Here are some ideas students might suggest.

    In the division machine, why is $c \div a$ bigger than c when a is less than 1 and a is positive?

    Why does $c \div a$ not keep moving to 0 or beyond as a moves farther from 0?

    Why does $c \div a$ keep moving to the left as c moves left into negative numbers? Why doesn't $c \div a$ bounce off 0 and go back to the right?

As a passes through 0, c ÷ a suddenly appears on the other end of the line. Is the line really a circle?

Why are there two places were a is right above c ÷ a if c is positive and no such place if c is negative?

ANSWERS

2. Marker *a* determines how many rectangles appear in the blue bar.

3. Marker *b* determines the width of each blue rectangle.

4. When *a* shows exactly one yellow square, marker *c* is exactly below marker *b* no matter where you drag it.

5. $c = 6$

6. $c = 15$

7. $c = -10$

8. You can calculate the length of the row of blue tiles by multiplying the number of tiles by the length of each tile. As an equation, $c = a \cdot b$.

9. When *a* is negative and *b* is positive, *c* is negative. When both are negative, *c* is positive. In all cases, the equation $c = a \cdot b$ remains true.

10. When you press the button, the control for *b* goes below the line and the control for *c* goes above it. Now you control *c* instead of *b*. This makes it a division machine that calculates $b = c \div a$.

11. $b = 5$

12. $b = 2.4$

13. $b = -3.6$

14. You would divide *c* by *a*: $b = c \div a$.

15. a. $b = 2.5$; drag *c* to 8.5; drag *a* to 3.4
    b. $a = 2.1$; drag *c* to 6.3; drag *a* until *b* equals 3
    c. $b = 2.1$; drag *c* to 10.5; drag *a* to 5
    d. $a = 4$; drag *c* to 15; drag *a* until *b* equals 3.75
    e. $c = 14.4$; drag *a* to 4; drag *c* until *b* equals 3.6

16. $c \div a$ moves in the same direction that *c* moves. They are equal when $c = 0$.

17. $c \div a$ moves in the opposite direction that a moves. $c \div a$ is equal to c when $a = 1$.

18. No. When you drag a and c so that both are negative, $c \div a$ is positive. Dividing by a negative number always changes the sign, so dividing a negative by another negative gives a positive answer.

19. No. When you drag a between 0 and 1, the result is that $c \div a$ is greater than c. Dividing a positive number by a number between 0 and 1 results in a greater positive number.

20. When you drag a toward 0, the marker for $c \div a$ moves off the screen, showing a greater and greater result. When you get a to exactly 0, $c \div a$ disappears entirely. Similarly, as you move a farther and farther away from 0, $c \div a$ moves closer and closer to 0, but never gets there.

22. To find 0, drag c until it is directly above $c \div a$. This spot is 0. This method works because 0 divided by any number (except 0) equals 0. To find 1, drag a (or c) until a and c are on top of each other. $c \div a$ is at 1. This method works because any number divided by itself equals 1. Alternatively, to find 1, drag a until c is directly above $c \div a$. The location of a is 1. This method works because dividing any number by 1 does not change the number.

Magic Dividing Machine

For GSP5   Name:

In this activity you'll explore division and how it relates to multiplication.

EXPLORE

1. Open **Magic Dividing Machine.gsp.** Go to page "Inverse Operations." You should be in Multiplication Mode. If not, press *Inverse Operation.*

2. Experiment by dragging markers *a* and *b* and observing what happens. How can you control the number of rectangles in the blue bar?

3. How can you control the width of the rectangles in the blue bar?

4. Drag *a* so there is exactly one yellow square. (If necessary, select the green marker and use the arrow keys on your keyboard.) How does *c* behave as you move *b*?

5. Press *Show Numbers.* Drag *a* to 2 and *b* to 3. (If necessary, press the buttons in the lower-right corner.) What is the value of *c*?

6. Drag *a* to 5. What is the value of *c*?

7. Drag *b* to −2. What is the value of *c*?

8. If you know the length of one blue tile and the number of tiles, you should be able to calculate the length of the row of blue tiles. How would you do this calculation? Write your answer as an equation using *a*, *b*, and *c*.

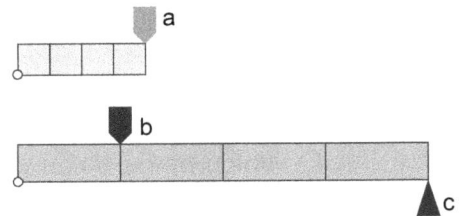

Magic Dividing Machine

continued

9. Experiment with other positive and negative values for both *a* and *b*. What happens when *a* is negative and *b* is positive? What happens when both are negative? Does your equation in step 8 always remain true?

10. Press *Inverse Operation*. What changed? Which markers can you control now? How is the machine different?

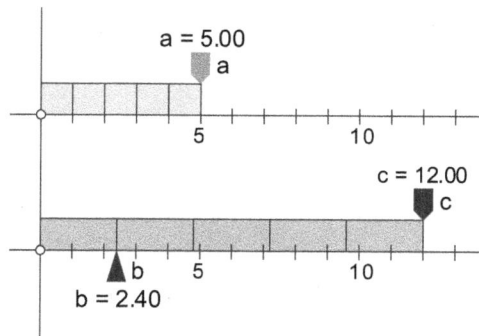

   a = 5.00
   a

   5   10

   c = 12.00
   c

11. Drag *c* to 15 and *a* to 3. What is the value of *b*?

   b
   b = 2.40   5   10

12. Drag *c* to 7.2. What is the value of *b*?

13. Drag *a* to −2. What is the value of *b*?

14. If you know the values of *a* and *c*, how would you use those numbers to calculate *b*?

15. While still in division mode, use the machine to solve the following problems. Explain how you dragged markers *a* and *c* to find each answer.

   a. $3.4 \cdot b = 8.5$
   b. $a \cdot 3 = 6.3$
   c. $10.5 \div 5 = b$
   d. $15 \div a = 3.75$
   e. $c \div 4 = 3.6$

New York City Title I Elementary School Activities Grades 1–5
© 2012 Key Curriculum Press

Magic Dividing Machine

continued

16. Go to page "Division Machine." Drag *c* left and right. How does *c* ÷ *a* behave? When is *c* ÷ *a* equal to *c*?

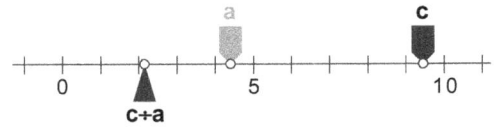

17. Drag *c* to a positive value, and then drag *a* left and right. How does *c* ÷ *a* behave now? When is *c* ÷ *a* equal to *c*? Press *Show Values* to check.

Use the division machine to answer the next three problems. For each question, tell how you dragged markers *a* and *b*, describe your observations, and explain your answers.

18. If you divide by a negative number, is the answer always negative?

19. If you divide by a positive number, is the answer always a smaller number?

20. What happens when you try to divide by 0?

EXPLORE MORE

21. Go to page "Mystery Machine." This is a division machine, but the numbers are left off. Drag *a* and *b*, and try to find the location of 0 and 1 on the unmarked number line. Drag the gold arrows for 0 and 1 to mark your answer. Then press *Show Answers* to check. Press *New Problem* to try the challenge several times.

22. Describe the strategies you used to find the locations of 0 and 1.

Jeff's Garden: Area Model of Fraction Multiplication

INTRODUCE

Project the sketch on a large-screen display for viewing by the class. Expect this part of the activity to take about 40 minutes.

1. Before using the sketch, pose the problem below. As you do, record the following for students to reference.

$\frac{3}{5}$ of the garden is Jeff's

$\frac{2}{3}$ of his part is for pumpkins

Jeff earns extra money by selling produce he grows in his grandmother's garden. This year, his grandmother will allow him to use $\frac{3}{5}$ of her whole garden. Jeff has decided that he will use $\frac{2}{3}$ of his part to grow pumpkins. What amount of his grandmother's garden will he use to grow pumpkins?

Some students may have learned an algorithm for multiplying fractions. Asking these students to make a drawing ensures that they think about making sense of the problem.

2. Provide paper and explain that students should use a drawing to show this situation. **When you come up with an answer, make sure you can explain why it makes sense.** Allow students to grapple with this, working in pairs or groups. As you circulate, ask questions to help students persist in reasoning about the problem. Because students often don't relate the word "of" to multiplying, don't expect them to think in terms of multiplying the fractions at this point.

What does it mean to have $\frac{3}{5}$ of a whole?

What could you show next in your drawing that would help you think about the problem?

Do you think Jeff is using more than, less than, or exactly half of the whole garden for pumpkin growing? Does your drawing make sense, given that idea?

3. Lead a discussion of students' drawings. Invite students' questions as well as their attempted representations. There is more than one way to draw this situation. Give students time to consider any different drawings that seem correct.

Becoming Familiar with the Model

4. Open **Jeff's Garden.gsp.** Go to page "Area Model." ***Let's see how we can use this Sketchpad model to represent the garden problem.*** Follow these steps.

To change a numerator or denominator, double-click the number. In the dialog box that appears, enter a new value.

- *The rectangle represents grandmother's garden, the whole garden.*

- *To represent Jeff's part, we need to show $\frac{3}{5}$ of the whole garden.* Change the denominator of the second fraction to 5. Students will see that the rectangle is now divided into fifths by vertical lines.

  Change the numerator to 3. Three of the fifths are now colored. Restate that this is Jeff's part of the garden, $\frac{3}{5}$ of the whole garden.

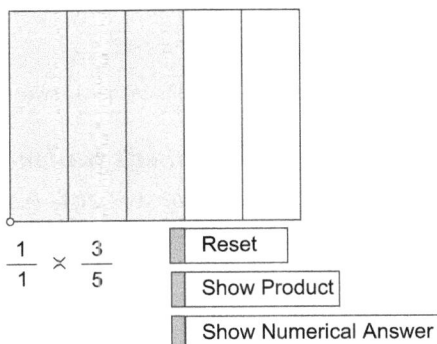

$$\frac{1}{1} \times \frac{3}{5}$$

Reset
Show Product
Show Numerical Answer

- *Now we want to show $\frac{2}{3}$ of Jeff's part. What would his part of the garden look like divided into thirds? Picture this in your mind.* Set the denominator of the first fraction to 3, and then drag the point *across the colored fifths only.* Students will see that two horizontal lines divide the colored area into thirds.

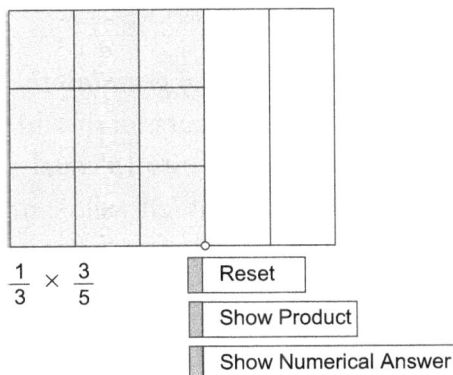

$$\frac{1}{3} \times \frac{3}{5}$$

Reset
Show Product
Show Numerical Answer

- *How many thirds of his garden will Jeff plant in pumpkins?* [Two] *Let's show $\frac{2}{3}$ of Jeff's part.* Change the numerator to 2 and press *Show Product.* Two of the horizontal regions are now colored. One of the thirds (the part not in pumpkins) is not.

$\frac{2}{3} \times \frac{3}{5}$

| Reset |
| Show Product |
| Show Numerical Answer |

• *What part of grandmother's garden—the whole garden—will Jeff plant in pumpkins?* Give students time to think, and then drag the point the remainder of the way across the rectangle.

$\frac{2}{3} \times \frac{3}{5}$

| Reset |
| Show Product |
| Show Numerical Answer |

Ask again, *What part of grandmother's whole garden will be planted in pumpkins?* The discussion should yield these ideas: Grandmother's garden is now divided into 15 equal parts; six of the parts (the colored region) are the amount Jeff will plant in pumpkins; so, Jeff will use $\frac{6}{15}$ of his grandmother's garden for pumpkins.

By asking whether the answer makes sense, you are modeling a question you want students to be asking themselves as they carry out computations.

• *Does this answer make sense?* Solicit thinking such as this student sample: *You can see that the garden is divided into fifteenths. Jeff's garden is 9 parts of the whole and 6 of those parts are for pumpkins. I know 6 is $\frac{2}{3}$ of 9, so $\frac{6}{15}$ makes sense for the $\frac{2}{3}$ of Jeff's garden.*

• Press *Show Numerical Answer* for confirmation of the answer.

$$\frac{2}{3} \times \frac{3}{5} = \frac{6}{15}$$

| Reset |
| Show Product |
| Hide Numerical Answer |

5. Provide a moment to reflect. *How does this model compare to the way you thought about the problem and the drawing you made at the start of this activity?* Take some responses, or have students discuss in pairs or small groups.

6. Model using the *Reset* button. Pressing the button removes the horizontal lines, the shaded area for the product, and the numerical answer. Students will need to set the fractions that are multiplied back to $\frac{1}{1}$ by changing the numerators and denominators themselves. Doing so will display an undivided rectangle.

DEVELOP

Expect students working at computers to take about 30 minutes.

7. Assign students to computers and tell them where to find **Jeff's Garden.gsp.** Distribute the worksheet. Explain that students should work on steps 1 and 2. *Using the model, make sense of the problems. Record your answers. Do the Explore More if you have time.*

8. As you circulate, listen for ways students are making sense of fraction multiplication. Here are some things to notice.

- If you need to help some students to reason as they work on Year 1 in the table, pose questions such as these: *Do you need to start with $\frac{3}{7}$? Suppose you started by thinking about $\frac{3}{7}$? What would $\frac{1}{3}$ of $\frac{1}{7}$ look like?*

The important thing here is that students not use the model, without thinking, merely following the steps they have learned.

- Notice any students who are making conjectures about other ways of solving fraction multiplication problems. If some students posit that the numerator in the product can be found by multiplying the numerators of the factors (and the same for the denominator in the product), ask, *Can you explain why that should work? Do you think it will work for every problem of this type?* Don't confirm for students at this point that this method will always work.

Jeff's Garden: Area Model of Fraction Multiplication

continued

Be on the lookout for any pairs who are using the model in an automatic way without making sense of each action they carry out. Ask, *What are you showing now? What part of the problem does that represent?*

- Some pairs may devise shortcuts for using the model. If you observe shortcuts being used, ask students to explain what they are doing and how they are thinking. If students are making sense of what they are doing, they should continue. For example, students may change *both* numerators and denominators to match the values in a problem before they drag the point and press *Show Product*. Or students may manipulate the model as demonstrated, but omit the last step of extending the horizontal lines across the whole garden. Most likely, they are able to mentally extend the lines and determine the pieces the whole rectangle is divided into.

SUMMARIZE

Project the sketch for viewing by the class. Expect to spend about 30 minutes.

9. Students should have their worksheets. Lead a discussion to explore the idea that multiplication by a fraction less than one results in a product smaller than the number being multiplied. (This is true for multiplying a whole number by a fraction and for multiplying a fraction by a fraction.) Begin by asking the following questions.

   You started with $\frac{3}{5}$ and multiplied it by $\frac{2}{3}$. Was the product (the amount planted in pumpkins) bigger or smaller than the number you started with, $\frac{3}{5}$?

   You multiplied and got a product smaller than the number you started with. Does that make sense?

   From their experiences with whole numbers, students assume that multiplication results in a product larger than the factors. Provide time for the class to grapple with this in order to develop a conceptual foundation that makes sense of fraction multiplication. Invite students to model at the computer as the class considers this issue.

It is important to spend time developing the idea that multiplying a fraction by a fraction is finding a fraction *of* a fraction.

   The discussion should bring out the idea that multiplying by a fraction involves taking *a part of* what you started with. Multiplying a fraction by a fraction involves *taking a part of a part*. For example, $\frac{1}{2} \times \frac{1}{5}$ means taking $\frac{1}{2}$ *of* $\frac{1}{5}$. You end up with less than you started with.

10. Discuss the problem in worksheet step 2. Note whether any students have related this problem to Jeff's garden in Year 2 (in worksheet step 1). In Year 2, Jeff plants $\frac{4}{5}$ of half the garden in pumpkins. In step 2, he plants half of $\frac{4}{5}$ of the garden in pumpkins. The answer is the

same in both cases, $\frac{2}{5}$. This suggests that multiplication of fractions is commutative. If students don't propose this idea, introduce it. Provide time for the class to explore one or two other problem pairs that are easy to visualize. *What is $\frac{1}{2}$ of $\frac{1}{4}$? What is $\frac{1}{4}$ of $\frac{1}{2}$?*

11. Present the following problem, which gives large values for the denominators. *Another year, Jeff's grandmother was not well. She gave him $\frac{7}{8}$ of the garden and kept a small strip for herself. Jeff planted $\frac{3}{15}$ of his area in green beans. What part of the whole garden did he plant in beans?*

The class is likely to anticipate that the model will display more parts in the solution than students wish to count. The intention here is to prompt them to look for strategies other than counting all the parts, if they haven't already. Provide a few minutes for students to work on the problem, recording on the back of their worksheets.

Ask students to share how they solved the problem. You may wish to say, *I saw some of you using shortcuts.* Call on students whom you noted using shortcuts, or ask for volunteers to model at the computer.

Some students are likely to have noticed that for all the garden problems, the product can be found by multiplying the numerators and multiplying the denominators of the factors. If this isn't suggested, ask, *How are the numerators and denominators represented in the model?* Set the model for $\frac{2}{3} \times \frac{3}{5}$, the problem you first modeled. Provide ample time for students to compare the symbolic representation with the area model.

$$\frac{2}{3} \times \frac{3}{5} = \frac{6}{15}$$

Reset

Show Product

Hide Numerical Answer

Invite explanations such as these student examples.

The colored area shows the numerators, $2 \times 3 = 6$, and the whole rectangle shows the denominators, $3 \times 5 = 15$.

Problems that come at the mathematics from a different angle are good for both strengthening and assessing student understanding.

The numerators are represented by two rows of three, and the denominators are represented by the three rows of five.

12. Discuss the Explore More problem, worksheet step 3. This is a working backward problem. Students know the part of the entire garden planted in pumpkins by Jeff and are asked to determine the part of Jeff's area that is planted in pumpkins.

EXTEND

1. Present problems with missing factors. Have students use the Sketchpad model to determine and/or check their answers.

2. Develop in students the habit of using computational estimation to anticipate and check computations for reasonableness. Present computations like these, and ask students to estimate whether the product in each computation will be greater or less than 1/2. Ask students to explain their reasoning.

$$\frac{5}{6} \times \frac{1}{3} \qquad \frac{4}{5} \times \frac{7}{8} \qquad 4 \times \frac{8}{9}$$

3. Have students write another problem like the Explore More. Students should exchange problems and solve. You might use this as an individual assessment.

ANSWERS

1.

Year 1	$\frac{3}{21}$
Year 2	$\frac{4}{10}$
Year 3	$\frac{21}{40}$
Year 4	$\frac{20}{72}$
Year 5	$\frac{10}{18}$

2. $\frac{4}{10}$, or $\frac{2}{5}$

3. $\frac{1}{2}$

Jeff's Garden

Name:

1. For the next five years, Jeff's grandmother gives him part of her garden. Jeff uses part of his area to grow pumpkins. What part of the whole garden does Jeff plant in pumpkins each year? Use **Jeff's Garden.gsp** to help you complete the table.

	Part of the Whole Garden Jeff Gets	Part of Jeff's Area He Plants in Pumpkins	Part of the Whole Garden Jeff Plants in Pumpkins
Year 1	$\frac{3}{7}$	$\frac{1}{3}$	
Year 2	$\frac{1}{2}$	$\frac{4}{5}$	
Year 3	$\frac{7}{10}$	$\frac{3}{4}$	
Year 4	$\frac{4}{9}$	$\frac{5}{8}$	
Year 5	$\frac{2}{3}$	$\frac{5}{6}$	

2. Another year, Jeff's grandmother gave him $\frac{4}{5}$ of the garden. He planted $\frac{1}{2}$ of it in pumpkins. What part of the whole garden did he plant in pumpkins? _____

EXPLORE MORE

3. One year, Jeff was given $\frac{2}{3}$ of the garden. He planted a part in pumpkins. His grandmother said, "This year $\frac{1}{3}$ of the whole garden is planted in pumpkins." Grandmother did not plant pumpkins. What part of Jeff's garden was planted in pumpkins? _____

New York City Title I Elementary School Activities Grades 1–5
© 2012 Key Curriculum Press

289

Function Machines: Introducing Functions ACTIVITY NOTES

INTRODUCE

Project the sketch for viewing by the class. Expect to spend about 20 minutes.

1. Open **Function Machines Introducing.gsp.** Go to page "Machine 1." Make sure the input showing is 2. To change an input, double-click it, enter a new value in the dialog box that appears, and click **OK.**

2. Explain, *This is an in-out machine. A number will go in when I press* **Run Machine.** *What number is about to go in?* [2] *What do you think will happen when I press* **Run Machine?** Take responses and then press the button. *What number came out of the machine? What do you think the machine did to change the "in" value of 2 to the "out" value, 4?*

   Encourage students' creativity. Students don't know yet that this is a one-step machine, so they may suggest possibilities such as *It multiplied the 2 by 4 and then divided by 2.* If students' ideas don't include the two possibilities that the machine added 2 to the input and that it multiplied the input by 2, offer these ideas. Record all suggestions on the board in the form, "multiply by 4, divide by 2" and "multiply by 2."

3. *Here's something else I can tell you about this machine: It only carries out* one *operation on the number that goes in. (We'll call that number the* input.*) For example, it might multiply the input. It doesn't do one operation, such as add, and then do another, such as multiply.*

   We'll call what the machine does its rule. *Now that you know that the machine has only one rule, which of the rules we've written down should we keep?* Have students identify "add 2" and "multiply by 2."

   Are you sure both of these rules could work for the input of 2 and output of 4? Though the answer may seem obvious, this is an important question; be sure to give all students enough time to consider it. *But the machine has only one rule! How do you think we can find out what that rule is?* Have students discuss this in pairs before taking responses.

   Some students may suggest trying another input, but they may not be ready to explain why that would be helpful. Other students may want to make the case that the class will need to test one or more new inputs because another input could give an output that would fit one of the rules, but not the other. *Give us an example of what you mean.* A sample student response is this: *Let's say we make the input 5. If the rule is "add 2," we know a 7 will come out of the machine. But if the rule is "multiply by 2," we know a 10 will come out of the machine.* **Who agrees with this idea?**

All students may not be convinced at this point. This is fine. The idea will be developed throughout the activity.

DEVELOP

Expect to spend about 25 minutes on this part of the activity. Continue to project the sketch.

4. Before the class embarks on trying other inputs and generating sets of input/output pairs, ask, *How might we keep track of the inputs we try and the outputs the machine produces?* Elicit the idea of making a table. Distribute the worksheet *In-Out Machines* and direct students' attention to Table A. Have students record 2 and 4, respectively, in the first row. In the sketch, create a table by selecting the input and output, in that order, and choosing **Number | Tabulate.** Double-click the table to enter the data.

It is possible to enter values for "*In*" that extend into the tenths place or beyond. However, the "*In*" and "*Out*" values on screen are set to display only to the nearest whole unit.

5. *Let's try another input.* Press *Reset* to prepare the machine to run again. Ask a volunteer to suggest a *whole number* for the input. Change the "*In*" parameter to this new value.

Ask students to predict the output for each of the rules the machine may be using: "add 2" and "multiply by 2."

Run the machine again. Have students record the pair in the second row of Table A. Double-click the table to add the resulting input/output pair.

6. *Now that you have another input and output, what do you think?* Facilitate discussion. Record ideas on the board and encourage students to debate them. Notice the ideas they have about whether they can predict with more certainty which of the two rules the machine is using. How many inputs do they think it's necessary to try in order to be certain of the rule the machine is using?

To save time, you can change the "*In*" value without first pressing *Reset.*

7. Run the machine several more times. Each time, encourage discussion about good numbers to use as inputs. Students may discover that small numbers, multiples of ten, and other "easy" numbers are simpler to work with than larger numbers. Students should record all input/output pairs in Table A as you double-click the Sketchpad table to add them.

8. When Table A is complete, ask students to study the inputs and outputs and write the rule they think the machine is using. Facilitate discussion. *Are you convinced you know what the rule is? Why or why not?* When most students are convinced they know the rule, press *Show/Hide Rule.* Does the rule shown, "add 2," confirm or challenge students' thinking?

New Machines, New Rules

9. Go to page "Machine 2" and facilitate as the class works to find the rule for this machine. The rule is now "subtract 3." Have students record inputs and outputs in Table B as you record in a Sketchpad table.

10. Go to page "Machine 3" and facilitate as the class works to find the rule for this machine. The rule is "multiply by 4." Have students record inputs and outputs in Table C as you record in a Sketchpad table.

SUMMARIZE

Expect to spend about 30 minutes on this part of the activity. Continue to project the sketch.

11. Have the class create problems of the type they have been solving—a good technique for helping them consolidate their understanding. ***Now you get to give the machines their rules! You can change the number associated with each rule, but not the operation.*** Using Machine 2, model these steps for changing a rule.

- Show the rule.

- Have a volunteer cover the projector lens.

- Have another volunteer change the value of the "*In*" parameter and press *Show/Hide Rule*.

Encourage creative challenges, such as adding 0 or multiplying by 0 or 1.

Have students make rules for the class to determine. Each rule maker can act as moderator at the computer, taking suggestions for inputs and running the machine. Students can create tables to keep track of input/output pairs.

12. Engage the class in shared writing to help students synthesize what they have experienced. ***Let's make a poster that tells what you've learned about in-out machines. What can you say about how in-out machines work? What can you say about determining the rule a machine is using?*** Here are three big ideas that should emerge from students' experience with in-out machines in this activity.

- If you know the rule a machine is using, for any input, you know what the output must be. The output for a rule depends on the input.

- Each input has a single corresponding output. If you run a machine again with the same input, the output has to be the same.

- To determine the rule, you probably need to know more than one input/output pair.

EXTEND

1. The work in this activity may have prompted new queries by students. Record the questions posed and provide time for students to investigate them. Here are some examples.

   How many numbers do we need to try before we know the rule for sure? Are two numbers enough?

   What if the rule were division?

   Could a machine have a rule other than add, subtract, multiply, or divide by a number? [For example, what if the machine squared the input?]

   Could a machine have two rules? How hard would it be to determine a two-step rule?

2. For students who would benefit from more individualized work with the sketch, consider giving them an opportunity to use **Function Machines Introducing.gsp** at a later time. Pairs of students can create and share in-out machine challenges with each other by changing the numbers associated with the rules.

3. Graph the input/output data you have collected in a table.

   - Select the table and choose **Edit | Copy.**

   - Add a blank page using **File | Document Options**.

   - Choose **Edit | Paste.** Hide any parameters that are showing.

   - Select the table and choose **Graph | Plot Table Data.** Click **OK.**

   - To see more points, drag the origin and change the scale of the grid.

In-Out Machines

Look for the rule an in-out machine is using.

Table A

In	Out

Rule: _____

Table B

In	Out

Rule: _____

Table C

In	Out

Rule: _____

Table D

In	Out

Rule: _____

New York City Title I Elementary School Activities Grades 1–5

Function Machines: Working Backward

INTRODUCE

Project the sketch for viewing by the class. Expect to spend about 15 minutes.

1. Distribute paper to each student and introduce the Sketchpad model. Open **Function Machines Working Backward.gsp** and go to page "Mystery Input 1." Explain, *A number we'll call the input is hidden behind a ball. When the input enters the machine, it will be multiplied by 3, and then 4 will be added to the result.* Press *Run the Machine.* The red ball will drop and an output of 13 will emerge. (The jiggling is a visual cue that the machine is working.)

2. *We don't know what input went into the machine. But we do know the output. Can you figure out the input?* Have students work on their own or in pairs for a few minutes to figure out the unknown input, and then have them share their methods. Here is a sample student response.

   I wasn't sure what the input could be, so I just picked a number and tried it. I started with 4, but that made an output of 16. That was too big, so I tried an input of 2. That was too small. I tried an input of 3—in between —and that worked.

Students' methods will become more refined in the Develop section that follows.

   Press *Show Input* to reveal the input.

DEVELOP

Continue to project the sketch. Expect to spend about 20 minutes.

3. Present problems with new mystery inputs in order to give students experience finding the inputs. For each problem, press *New Input.* (The computer will assign a random value from 0 to 12 for the input.) Run the machine again. Ask students to figure out the value of the input. (It's possible that the sketch will pick the same input twice. If it does, simply press *New Input* again.)

   As students solve the problems, note whether anyone is working backward.

   I thought I could go backward. The output this time was 10. That means some number became 10 when 4 was added to it. That must be 6. Six is some number multiplied by 3. That number has to be 2. So, the input should be 2. I checked by seeing whether 2 times 3, plus 4, equals 10. It does!

   If this method is not proposed, ask guiding questions. *How can we figure out what number entered the "add" machine?* Elicit the idea that, because the machine added 4 to some number to obtain the final output, we need to subtract 4 from the output to find out what went into the "add" machine.

Continue this reasoning by asking about the first machine. ***How can we figure out the "multiply" machine's input now that we know its output?*** Because the input was multiplied by 3 to get the output, the output must be divided by 3 to find the input.

Let's check. Imagine the number you think is the input goes into the machine. Apply the two-step rule. Do you get the output?

4. Ask students to describe the two-step rule for finding the input when given the output. For the example above, the rule is "subtract 4, divide by 3."

Solving Problems with Different Rules

5. Now create problems by changing the numbers associated with the two-step rule. Double-click a number with the **Arrow** tool, enter a new number in the dialog box, and click OK. In this manner you can change the "multiply by 3, add 4" rule to another rule, such as "multiply by 2, add 5." (Note that you can change only the numbers, not the operations.)

6. Have students solve problems with different operations by using the machines on pages "Mystery Input 2" and "Mystery Input 3." Have students describe a two-step rule for finding the input. (Note that the "Mystery Input 3" machine displays outputs to the nearest tenth. This works fine for "divide by 2," but will be problematic if you choose a rule like "divide by 3.")

SUMMARIZE

Expect to spend about 20 minutes.

7. Ask what students notice about how they have determined the inputs. Here are sample student responses.

We used the opposite operations to undo what the machine was doing. When it added, we subtracted. When it multiplied, we divided.

Our rules are upside-down from the machine's rules. Say the machine starts with multiplying and ends with adding. We start with subtracting and end with dividing.

Elicit the ideas that (1) addition and subtraction "undo" each other, and (2) the strategy students have been using is called working backward, or "undoing."

8. Have students create their own missing input problem to share with friends or family. Distribute the worksheet *Mystery Input* and have students enter their own two-step rule in the blank spaces next to the machine. They should also enter several outputs into the accompanying input/output table, leaving the input column blank.

   Discuss with students how they will describe the missing input problem when presenting it to a friend or family member. Explain that students should not offer strategies for solving the problem at first.

   Have the class report back on the solution strategies they observed others use to solve the problem. Highlight students' accounts of the working backward (or "undoing") strategy.

EXTEND

1. For students who would benefit from more individualized work on undoing function rules, provide opportunities to work with any of the three Mystery Input machines.

2. Pairs of students can also use the three machines to create challenges for each other. As one student looks away, the other student presses *Show Input.* She double-clicks the value of the input, enters a new value in the dialog box, and clicks OK. Finally, she presses *Hide Input* to conceal the input, and challenges her partner to determine it.

Mystery Input

For GSP5

Name:

Make a machine for someone else.

1. Write a rule such as "add 4" or "multiply by 3" in each of the blanks.

2. Choose some inputs and find their outputs. Write *only* the outputs in the table. (Try to use mental math to do this.)

3. Share this problem with a friend or family member. Can he or she figure out the inputs?

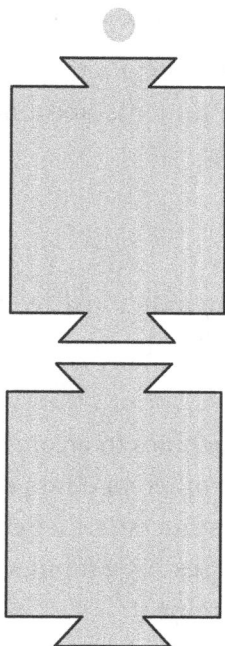

Input	Output

Perfect Packages:
Surface Area and Volume

For GSP5

ACTIVITY NOTES

INTRODUCE

Project the sketch for viewing by the class. Expect to spend about 10 minutes.

1. Open **Perfect Packages.gsp.** Go to page "Box and Net."

2. Explain, *Today you'll look at two different ways of describing the size of a package that is a rectangular prism. On the left side of the sketch you see a package that has a volume of 6 cubic units. On the right side this box has been unfolded into a net and the area of the net corresponds to the surface area of the package. What is the surface area of this package?* Allow students to propose answers and encourage them to describe how they calculated their answers. It may be useful to point to the different-colored rectangles on the net and relate them to the sides of the package. *As you can see, the surface area and the volume are not usually the same. Sometimes, depending on the purpose of the package, people want to make a box that uses as little surface area as possible while having a certain volume. Can you think of an example of when this might happen?*

3. Show students how to drag the sliders to change the length, width, and height. Ask, *Can someone tell me how I could make a package that has the same volume as this one, but a different shape?* Use the sketch to illustrate students' examples. These might include 1 × 1 × 6. Ask students whether a 3 × 1 × 2 package should be considered to have the same shape as a 2 × 3 × 1 package. Opinions may differ, but let students know that this activity assumes they are the same shape. Ask, *Do you think that all the packages with a volume of 6 cubic units have the same surface area? This is the first question you'll be exploring today.*

DEVELOP

Expect students at computers to spend about 25 minutes.

4. Assign students to computers and tell them where to locate **Perfect Packages.gsp.** Distribute the worksheet. Tell students to work through step 7 and do the Explore More if they have time.

5. Let pairs work at their own pace. As you circulate, here are some things to notice.

   • Help students find shorter ways than counting to calculate the surface area of the box.

   • Help students develop strategies to make sure they have found all possible ways of creating a box with a given volume. This might involve making a table or listing all the factors of the value of the volume.

- Encourage students who finish quickly to investigate dimensions that are not whole number values (in the Explore More).

SUMMARIZE

Project the sketch. Expect to spend about 10 minutes.

6. Open **Perfect Packages.gsp** and use it to support the discussion. Invite students to share their strategies for finding the surface area of a package. If students bring it up, you might add to the projected sketch the surface area calculation described in the Extend section.

7. Ask students to talk about which packages had the least and greatest surface areas for a given volume. Students should notice that the least surface area occurs when the dimensions of the package are as equal as possible (that is, when the package is as cubic as possible), and that the greatest surface area occurs when two of the dimensions are very small and the third dimension is very large (that is, when the package is as long and thin as possible). Encourage students to consider dimensions that are not whole numbers. In this case the least surface area will occur when the dimensions are all the cube root of the volume.

EXTEND

1. Use Sketchpad's calculator to write a general equation for the surface area of the package (such as $2*length + 2*width + 2*height$, or $2*(length + height + width)$). Change the value of one of the dimensions and investigate how both the volume and the surface area change.

2. Discuss how the relationship between volume and surface area might change for prisms that don't have right angles, or for other solids. If students already know that a circle gives the maximum area for a given perimeter, you can discuss how a sphere gives maximum volume for a given surface area.

ANSWERS

1. Surface area = 22 square units

2. $1 \times 1 \times 6$ box: volume = 6 cubic units; surface area = 26 square units

   $1 \times 1 \times 5$ box: volume = 5 cubic units; surface area = 22 square units

3. The values for length, width, and height are interchangeable.

   $2 \times 3 \times 4$ arrangement: surface area = 52 square units

   $2 \times 2 \times 6$ arrangement: surface area = 56 square units

   $1 \times 4 \times 6$ arrangement: surface area = 68 square units

   $1 \times 3 \times 8$ arrangement: surface area = 70 square units

   $1 \times 2 \times 12$ arrangement: surface area = 76 square units

   $1 \times 1 \times 24$ arrangement: surface area = 98 square units

4. The $2 \times 3 \times 4$ arrangement has the least surface area. The $1 \times 1 \times 24$ arrangement has the greatest surface area. (On page "Decimal Values," using approximately 2.9 units for each dimension will produce the least surface area.)

5. a. $2 \times 2 \times 2$ (surface area = 24 square units)
   b. $2 \times 2 \times 3$ (surface area = 32 square units)
   c. $3 \times 3 \times 3$ (surface area = 54 square units)
   d. $3 \times 4 \times 4$ (surface area = 80 square units)

6. a. $1 \times 1 \times 8$ (surface area = 34 square units)
   b. $1 \times 1 \times 12$ (surface area = 50 square units)
   c. $1 \times 1 \times 27$ (surface area = 110 square units)
   d. $1 \times 1 \times 48$ (surface area = 194 square units)

7. To find the least surface area, make the dimensions as equal as possible (so the package is as close to a cube as possible). To find the greatest surface area, make two of the dimensions equal to 1, and the third dimension equal to the volume (to make as long and thin a package as possible).

8. The package with the least surface area for 5b (volume = 12) is when each dimension is approximately 2.3 units. The package with the least surface area for 5d (volume = 48) is when each dimension is approximately 3.6 units.

The greatest surface area for each volume is when two dimensions are 0.1 unit and the third dimension is very large, but these no longer represent realistic packages.

9. Answers will vary. Sample answers: $1 \times 1 \times 5.5$ (volume = 5.5 cubic units); $1 \times 1.6 \times 4$ (volume = 6.4 cubic units); $2 \times 2 \times 2$ (volume = 8 cubic units)

Perfect Packages

For GSP5   Name:

In this activity you'll investigate how changing the dimensions of a package (a rectangular prism) affects volume and surface area. Of course, you'll be most interested in the biggest and smallest packages you can make!

EXPLORE

1. Open **Perfect Packages.gsp** and go to page "Box and Net."

   If needed, change the dimensions to match those pictured at right, so the volume is 6 cubic units. What is the surface area?

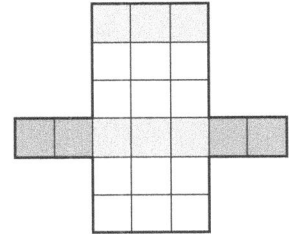

2. Can you find a package that has the same volume, but a different surface area? Can you find a package with the same surface area, but different volume? Drag the length, width, and height sliders to see whether you can come up with examples of such packages. Describe what you found.

3. Find all the ways 24 cubes can be arranged into a package. For each arrangement, write the dimensions and surface area in the table.

Length	Width	Height	Volume	Surface Area
			24 cubic units	
			24 cubic units	
			24 cubic units	
			24 cubic units	
			24 cubic units	
			24 cubic units	
			24 cubic units	
			24 cubic units	

4. Which of your arrangements has the least surface area? Which has the greatest?

5. For each given volume, which package has the least surface area?
   a. Volume = 8 cubic units
   b. Volume = 12 cubic units
   c. Volume = 27 cubic units
   d. Volume = 48 cubic units

6. For each given volume, which package has the greatest surface area?
   a. Volume = 8 cubic units
   b. Volume = 12 cubic units
   c. Volume = 27 cubic units
   d. Volume = 48 cubic units

7. For any given volume, describe how you would find the packages with the least and greatest surface areas.

EXPLORE MORE

8. Go to page "Decimal Values," where you can adjust the dimensions to the nearest tenth of a unit. See whether you can find packages whose surface areas are less than those you found in step 5, or greater than those you found in step 6. Describe what you found.

9. Suppose you have a surface area of 24 square units. Can you make packages with different volumes? If so, provide two possibilities.

Prism Dissection: Surface Area

INTRODUCE

Project the sketch for viewing by the class. Expect to spend about 5 minutes.

1. Show students the sample full-page net you cut out, and fold it to form a regular pentagonal prism. Explain, *Today you will find the surface area of regular right prisms. You'll start with a pentagonal prism and then figure out a formula that works for other regular right prisms.*

2. Open **Prism Dissection.gsp.** Explain, *On page "Prism," you'll review how to change the viewing angle and how to change the dimensions and number of sides.* Model the use of the *spin, pitch,* and *roll* controls and how to change the number of sides.

3. *What does surface area mean?* Review the definition with the class. Here is a sample definition: The surface area of a prism is the combined area of all the faces of the prism. *If you want to find the surface area of a prism, how can a net help you?* Encourage students to explain why the area of the net is the same as the surface area of the prism itself.

4. If students have not previously used the Construct or Measure menus, briefly demonstrate how to construct a midpoint and how to measure the distance between two points. If students have not previously used the Calculator, show them how to choose **Number | Calculate** and how to click an object in the sketch to enter it into a calculation.

DEVELOP

Expect students at computers to spend about 30 minutes.

5. Assign students to computers and tell them where to locate **Prism Dissection.gsp.** Distribute the worksheet. Tell them, *Once you've reviewed the controls on page "Prism," your next job will be to go to page "Base 1" and find the area of one base. If you can, try to figure out the measurements and calculations you need without using the hint.*

6. Give students time to work on their measurements and calculations in worksheet steps 5 and 6. If necessary, remind them to enter existing measurements into the Calculator by clicking on them rather than by typing the numbers in. Some students may be ready to go on to page "Base 2" before others have finished their calculations. If they do so, you can check their results for step 6 to make sure they answered 3.63. (All students should get this same result for a pentagonal base, no matter what size it is.)

7. When most students have finished worksheet step 6, call the class to attention and ask several students to report their values of r, the length of a side, and the step 6 calculation. Ask, ***What's interesting about these results from different-size bases?*** There's no need at this point for students to explain why the ratios are all the same, but it is important to bring this fact to their attention.

8. Tell students, ***If you used the number 5 in your calculation of the area, your calculation will work only for bases that are pentagons. On page "Base 2," do the same calculation, but in a way that will work for any polygon, no matter how many sides it has.***

9. Give students time to work on worksheet steps 7–11. Some students may be ready to go on to page "Faces" before others have finished their calculations. If they do so, you can check their results for step 10 to make sure they answered 3.14.

10. When most students have finished worksheet step 11, call the class to attention and ask several students to report their values of r, n, and the step 10 calculation. Ask, ***What's interesting about the calculated ratio now?*** Some students may mention that not only are the ratios very close to equal, but they get closer and closer to π as the number of sides gets larger and larger. Don't discuss this result in detail yet; give students time to think about it while they do the next few steps of the worksheet.

11. Tell students, ***Now that you've calculated the area of a base, you also need to calculate the area of a lateral face of the prism. Go on to page "Faces" and then to page "Area" when you're ready. If you finish early, try some of the Explore More questions.***

12. Give students time to work on worksheet steps 12–16. You might also consider assigning some or all of the Explore More, depending on your curriculum needs. If so, allot additional time.

SUMMARIZE

Project the sketch.
Expect to spend about
10 minutes.

13. Have students discuss their results. Here are some questions for discussion.

How did you find the area of the base?

How did the net help you to understand the problem of finding the surface area?

You ended up with a formula that involved the values of n, r, h, and the length of a side of the polygon. Do you really need all of these values?

Encourage students to realize that these are interdependent. At the middle school level, you might discuss that for a given number of sides, the side length increases in proportion to the value of r. As you drag R, observe the side length and note that the triangles used to find the area of the base remain similar. At the high school level, you can use trigonometry to explicitly relate the side length to the values of r and n.

What interesting things happen as n increases? Students will observe that the shape becomes more and more like a cylinder. *What happens to the area of the base?* Some students will have noticed that the ratio in worksheet step 10 is 3.14, and that as the number of sides increases, the ratio approaches π. Some also may have drawn the conclusion that the area of the polygon approaches πr^2, the area of a circle. *What happens to the area of the lateral faces?* Some students will have noticed that the ratio in worksheet step 15 approaches 2π. Some may also have drawn the conclusion that the area of the lateral faces approaches $2\pi rh$, the circumference of the circle multiplied by the height. (Explore More worksheet step 18 explicitly asks students to use these facts to develop a formula for the surface area of a cylinder.)

ANSWERS

3. Answers will vary.

For the area calculations in steps 5 and 8, students can partition the base into 5 triangles with bases that are the side length (represented by s here) or 10 triangles with bases that are half the side length (represented by t here).

5. $A = 5 \cdot (s \cdot r)/2$ (alternatively, $A = 10 \cdot (t \cdot r)/2$)

6. For a five-sided polygon, the ratio is 3.63 and does not depend on the size.

8. $A = n \cdot (s \cdot r)/2$ (alternatively, $A = 2n \cdot (t \cdot r)/2$)

9. The value of the ratio depends on n. Any two students using the same value of n should have the same ratio.

10. For approximately $n = 100$, the ratio is 3.14. As n becomes large, the ratio appears to approach π. This makes sense because the more sides the polygon has, the more it looks like a circle, and the closer the measurement r is to the radius of the circle. Because the area of a circle is πr^2, the ratio of the area to r^2 is π.

11. It makes sense to use r because r measures the radius of the limiting circle. In fact, when n is small, r is the radius of the inscribed circle, though students may not make this observation specifically.

13. The area of the lateral faces is $n \cdot s \cdot h$.

14. When you divide by $r \cdot h$, the resulting value changes with n, but not with r or h.

15. When the value of n is large, the ratio becomes approximately 2π. One explanation is that the face area is equal to the perimeter $(n \cdot s)$ multiplied by h. As n increases, the perimeter approaches the circumference of a circle, so the face area approaches $2\pi r$ multiplied by h. When this value is divided by rh, the result is 2π.

16. The total surface area of the prism is the sum of the base areas and lateral face areas, so it can be expressed as $A = 2(n(s \cdot r)/2) + n(s \cdot h) = n \cdot s \cdot r + n \cdot s \cdot h$. The product $n \cdot s$ is the perimeter of the base (p), so the formula could be written as $n = p \cdot r + p \cdot h$.

18. When n becomes large, $n \cdot s$ approaches $2\pi r$, so the area approaches $2\pi r^2 + 2\pi rh$.

19. The volume must equal the area of the base multiplied by the height: $V = n \cdot s \cdot r \cdot h$, approaching a limit of $V = \pi r^2 h$.

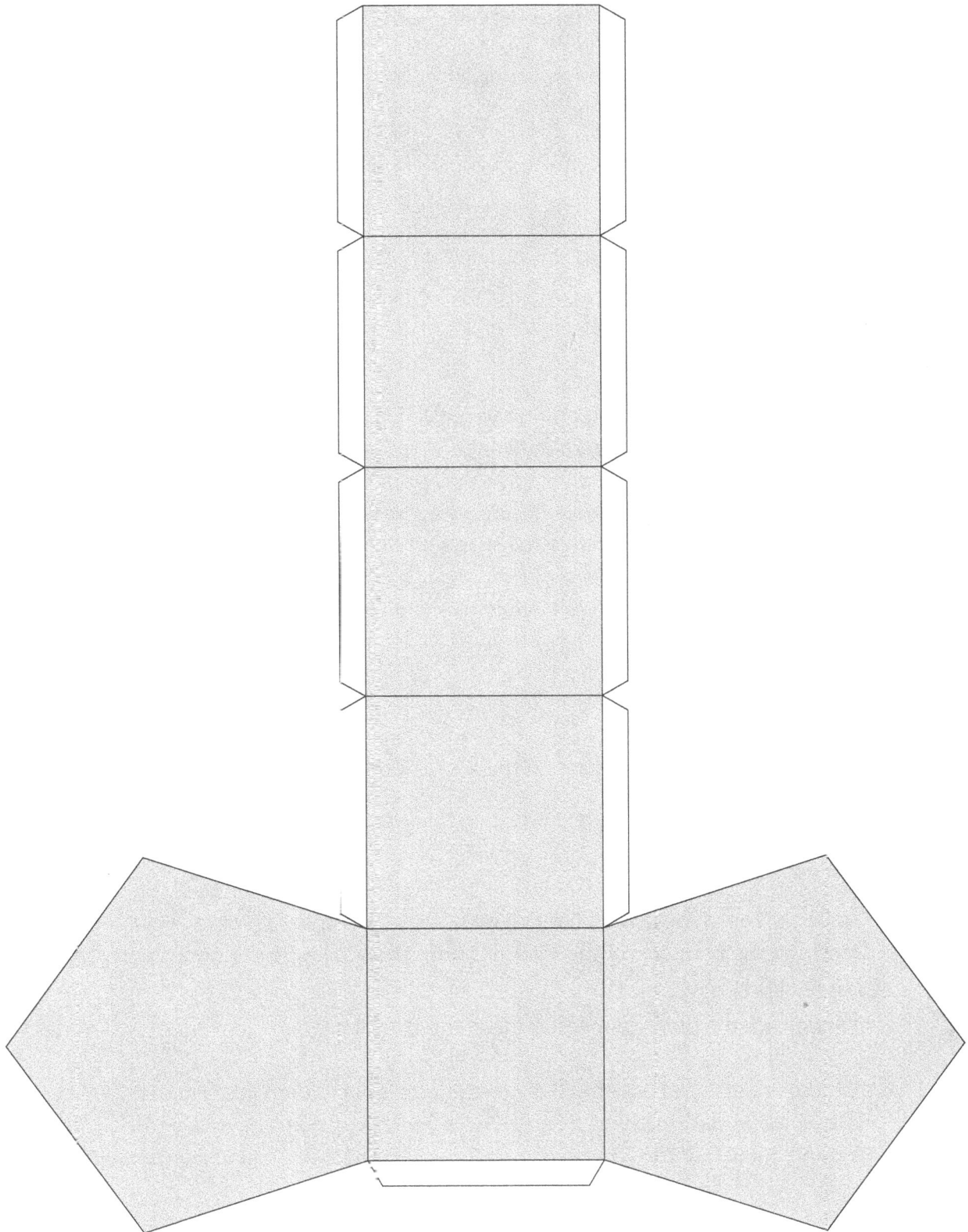

Prism Dissection

For GSP5 Name:

In this activity you will create a regular prism with your choice of the height, the number of sides, and the size of the base. Then you'll calculate its surface area.

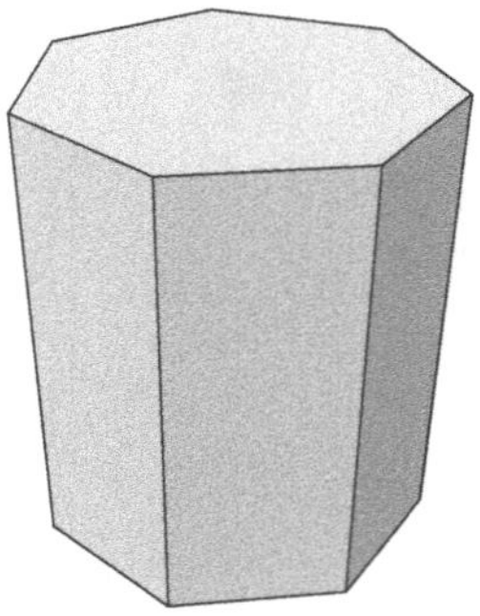

EXPLORE

1. Open **Prism Dissection.gsp** and go to page "Prism." Drag *spin, pitch,* and *roll* to view the regular prism from different angles.

2. Drag *N* to increase the number of sides. View the new prism from different angles. Drag *R* and *H* to change the shape of the prism.

3. How does the two-dimensional net correspond to the shape of the prism?

To find the surface area, you must find the area of both bases and of the lateral faces.

Base Area

4. On page "Base 1," the base has five sides. Change the size by dragging *R.* Construct the midpoint of the thick red side. Then measure the distance from *R* to the midpoint.

   Double-click the distance measurement and label it *r*.

5. Find the area of the base. You'll divide the regular pentagon into simpler shapes, measure some distances, and do some calculations. (If you're stuck, pressing *Show Hint* can give you some ideas.) Write down your calculation and result.

6. Divide your resulting area by r^2. What value do you get? How does it compare with the results of other students?

7. On page "Base 2," measure distance r as you did in step 4. Change the number of sides and the size of the base.

8. Using n to represent the number of sides, write a calculation for the area of the base that will be correct for any value of n. Write down the result for your value of n.

9. Divide the area by r^2. What value do you get?

10. Increase the number of sides to more than 50. Now what is the value of the area divided by r^2? Have you seen this number before? Why do you think you get this value?

11. Why does it make sense to use the letter r to stand for the distance from the center to a side of the base?

Face Area

12. On page "Faces," drag H, R, and N to change the shape of the prism and the number of sides. Measure the distance r as you did in step 4. Then measure the height of the prism, and label it h.

13. Find the total area of the lateral faces. Write a calculation that will be correct for any value of *n*. Write down the result for your value of *n*.

14. Divide your resulting area by the product of *r* and *h*. What value do you get? Does this value change if you drag *R* or *H*? Does it change if you drag *N*?

15. What is this value when the number of sides is at least 50? Why?

Total Area

16. On page "Area," do the necessary constructions, measurements, and calculations to find the total area of the regular prism, including both bases and the lateral faces. Write down your calculation and result.

EXPLORE MORE

17. On page "Net," set *H*, *R*, and *N* to match the prism you made on page "Area." Choose **File | Page Setup**, set the page to print in landscape view, and then choose **File | Print Preview**. If necessary, change the scale so the net fits on one page. Then click **Print**. Cut out the net, fold along the lines, and glue or tape your three-dimensional prism together. Label each base and face with the area you calculated based on the measurements.

18. What does the prism look like when the number of sides is very large? How could you calculate the approximate surface area of this shape without using the value of *n*? (Your answers to steps 10 and 15 may be useful.)

Prism Dissection

continued

19. Use your measurements to calculate the volume of the prism. Explain why you used the calculation you c d. Then write your method as a formula.

20. On page "Explore More," you can experiment with the advanced controls that affect the look of this three-dimensional model.

Pyramid Dissection: Surface Area

INTRODUCE

Project the sketch for viewing by the class. Expect to spend about 5 minutes.

1. Students should have already completed the activities Prism Nets and Prism Dissection. Show students the sample full-page net you cut out, and fold it to form a regular pentagonal pyramid. Explain, *Today you'll use Sketchpad to find the surface area of regular pyramids. You'll start with a pentagonal pyramid and then figure out a formula that works for other regular pyramids.*

2. Open **Pyramid Dissection.gsp.** Explain, *On page "Pyramid," you'll review how to change the viewing angle and how to change the dimensions and number of sides.* If students are not already familiar with the controls, model the use of *spin, pitch,* and *roll,* and the use of *N, R,* and *L.*

3. *To find the surface area of a pyramid, what would you have to measure?* Some students might provide a general description that you must measure the areas of the base and the five triangular lateral faces. Others might give more detail, describing how to measure the base and height of each triangle. *How can a net help you?* Students should see that the area of the net is equal to the surface area of the pyramid.

4. Holding up the folded pyramid, ask, *Where would you measure the height of this pyramid?* Encourage students to notice that there are two possible height measurements: the vertical distance from the center of the base to the vertex (height of the pyramid) and the distance from the base of a lateral face to the vertex (height of a triangle). Encourage students to discuss how these two different measurements might be useful, and discuss with them why it's important to avoid confusion by distinguishing the two heights. You may want to ask them to propose their own names for these measurements. Explain, *On your worksheet, the distance from the base of a triangular face to the vertex is called the* **slant height** *and is labeled l.*

5. Students should have previous experience with the Construct and Measure menus. You may want to briefly review how to construct a midpoint, how to measure the distance between two points, how to change the label of a measurement, and how to click on an object in the sketch to enter it into the Calculator.

DEVELOP

Expect students at computers to spend about 30 minutes.

6. Assign students to computers and tell them where to locate **Pyramid Dissection.gsp.** Distribute the worksheet. Tell them, *First review the controls on page "Pyramid" and answer the questions in steps 3 and 4. Then you'll go to page "Base" and find the area of the base. Try to figure out the measurements and calculations you need without using the hint.*

7. Give students time to work on their measurements and calculations. If necessary, remind them to enter existing measurements into the Calculator by clicking on them rather than by typing in the numbers. Some students may be ready to go on to page "Faces" before others have finished their calculations. If they do so, you can check their results for worksheet steps 6, 7, and 9. For step 9, make sure they answered approximately 6.29 for the ratio of the perimeter to r, and approximately 3.14 for the ratio of the area to r^2. Also make sure they've written an explanation for why they got these values.

   If students have different answers for worksheet step 9, check to make sure that they used the value of n and the actual measurements rather than typing in numbers, so that their results are correct no matter how they change the number of sides.

8. When most students have finished worksheet step 9, call the class together and ask several students to report their measurements and calculations of perimeters and areas from worksheet steps 6 and 7. Ask, *What values did you get for the two calculations in step 9?* Encourage students to explain why the two ratios come out to 2π and π. You may want to ask them whether the values are exactly 2π and π, and if not, why not? How could they measure the percentage by which the ratios differ from 2π and π?

9. Ask students, *When you did your calculations, why was it important to click on the measurements in the sketch rather than just typing in the number?* Encourage students to observe that they want their calculations to be correct even when they change the pyramid dimensions or the number of sides.

10. Tell students, *Now you'll calculate the area of the lateral faces of the pyramid. Go on to page "Faces" and then to page "Area" when you're ready. If you finish early, try some of the Explore More problems.*

11. Give students time to work on worksheet steps 10–13. In step 12, some students may need a hint; you can suggest that they see what happens if they divide their numeric answer by *rl*.

    Consider assigning some or all of the Explore More questions, depending on your curriculum needs. If you do so, allot additional time.

SUMMARIZE

Project the sketch. Expect to spend about 10 minutes.

12. Have students discuss their results. Here are some questions for discussion.

    How did you find the area of the base?

    How did the net help you to understand the problem of finding the surface area?

    You ended up with a formula that involved the values of n, r, l, and the length of a side of the base. Do you really need all of these values?

    Encourage students to realize that these are interdependent. At the middle school level, you might discuss that for a given number of sides, the side length increases in proportion to the value of *r*. As you drag *R,* observe the side length and note that the triangles used to find the area of the base remain similar. At the high school level, you can use trigonometry to relate the side length explicitly to the values of *r* and *n*.

13. *What interesting things happen as n increases?* Students will observe that the shape becomes more and more like a cone. *What happens to the area of the base?* Students will have noticed that the ratio of area to r^2 in worksheet step 9 approaches π. They should be ready to draw the conclusion that the area of the polygon approaches πr^2, the area of a circle. *What happens to the total area of the lateral faces?* Some students will have noticed that the ratio of the perimeter to *r* in worksheet step 9 approaches 2π. Encourage students to explain how they could use the perimeter to make it easy to calculate the sum of the areas of these faces, by calculating *perimeter × slant height*/2. (Explore More worksheet step 15 explicitly asks students to use these facts to develop a formula for the surface area of a cone.)

EXTEND

What other questions might you ask about pyramids? Encourage all inquiry. Here are some ideas students might suggest.

What do pitch, roll, and spin really do? Are there other similar movements?

Why are the circumference and the area of a circle given by the familiar formulas? Is this activity a proof?

Can you do something like this to find the volume of the pyramid or cone?

Can you get the surface area of other curved figures, like a sphere, by taking some figure and increasing the number of faces?

ANSWERS

3. Student answers will vary. The *pitch* control allows you to see a top view, so that the pyramid looks like a regular polygon. In this view the *spin* control rotates the polygon about its center.

4. You cannot make the slant height *l* smaller than the value of *r* because the slant height must reach at least from the edge of the base to the center. When the slant height is equal to *r*, the pyramid is completely flat, with a height of 0. When the slant height is large compared to the radius, the pyramid is tall and skinny.

6. If students use *s* for the length of one side of the base, the perimeter is *ns*. Numeric results will vary and should change as students manipulate the dimensions.

7. One way to measure the area of the base is to think of it consisting of *n* triangles (as suggested by the hint), to measure the base (*s*) and height (*r*) of one triangle, and then use $A = sr/2$ to find its area. To find the area of the entire base, students can multiply by *n*, with the result that the total area is given by $nsr/2$.

9. The ratio of the perimeter to *r* is approximately 6.29—nearly 2π—and the ratio of the area to r^2 is approximately 3.14—nearly π. When $n = 60$, these values are accurate to about two decimal places.

11. The area of one lateral face is $sl/2$, and the sum of the areas of all these faces is $nsl/2$. If students use the perimeter *p* in their calculations, they will use the formula $pl/2$. Numeric answers will vary.

12. When the number of sides is large ($n \geq 50$), the perimeter approaches $2\pi r$, so the area of the sides approaches πrl.

13. The total area of the pyramid is the sum of the area of the base and the lateral faces: $nsr/2 + nsl/2$. Students may factor this to write it as $ns(r + l)/2$. Numeric results will vary.

15. When the value of n is large, the base becomes very nearly a circle, and the base area can be written as πr^2. The sum of the areas of the lateral faces approaches πrl, so the surface area of a cone is given by $\pi r^2 + \pi rl$, or $\pi r(r + l)$.

16. The height h of the pyramid (measured from the center of the base to the vertex) and the distance r form two legs of a right triangle, with the slant height l forming the hypotenuse. By the Pythagorean Theorem, $r^2 + h^2 = l^2$. If $l = 15$ cm and $r = 9$ cm, $h = 12$ cm. If $h = 5$ cm and $r = 12$ cm, $l = 13$ cm.

New York City Title I Elementary School Activities Grades 1–5
© 2012 Key Curriculum Press

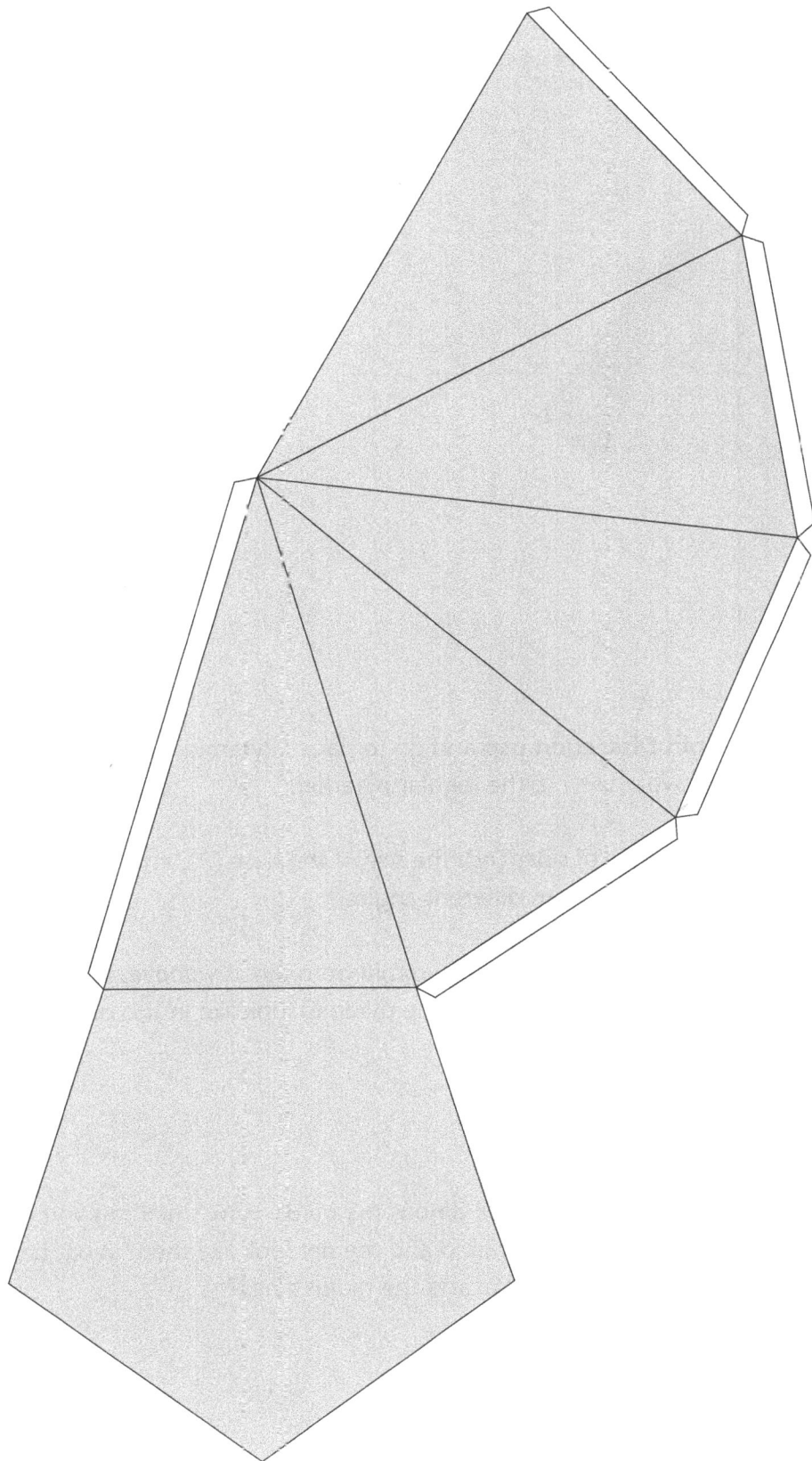

Pyramid Dissection

Name:

In this activity you'll create a regular pyramid with your choice of the height, the number of sides, and the size of the base. Then you'll calculate its surface area.

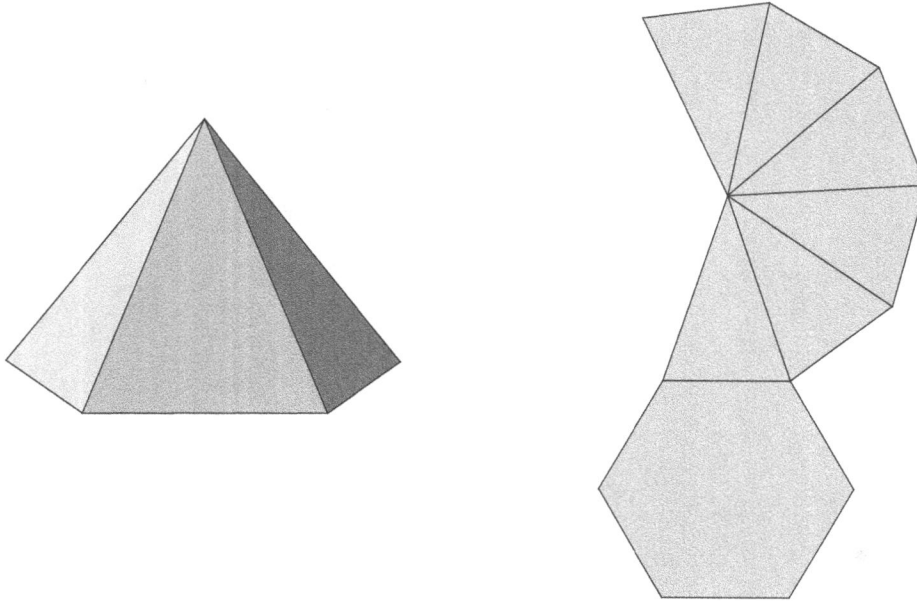

EXPLORE

1. Open **Pyramid Dissection.gsp** and go to page "Pyramid." Use *spin, pitch,* and *roll* to change your view of the regular pyramid.

2. Change the number of sides (*N*), the size of the base (*R*), and the slant height (*L*). View the pyramid from different angles.

3. Adjust the controls to look at the pyramid from directly above. Which control did you use to do this? What does the pyramid look like in this position? What does the *spin* control do now?

4. Explore the shape of the net for various pyramids. How small can you make the slant height? What do the pyramid and the net look like then? What happens if you make the slant height large and the radius small?

Pyramid Dissection

continued

To find the surface area, you must find the area of the base and of the lateral faces.

Base Area

5. Go to page "Base." Change the size of the base and change the number of sides. Measure the distance from R to the midpoint of the thick red side. Label the measurement r.

6. Measure one side of the base and have Sketchpad calculate the perimeter. Write down your calculation and result.

7. Find the area of the base. Imagine dividing the regular polygon into simpler shapes and do some measurements and calculations. (If you're stuck, press *Show Hint* to get some ideas.) Write down your calculation and result.

8. Change the number of sides and the size of the base, and make sure that your perimeter and area calculations seem reasonable. If not, fix them so that they work correctly for any base.

9. Increase the number of sides to more than 50. Divide the perimeter by r and divide the area by r^2. Have you seen these two numbers before? Why do you think you get these values?

Face Area

10. On page "Faces," drag L, R, and N to change the slant height, the size of the base, and the number of sides. Measure the distance r and side length s as you did in steps 5 and 6. Measure the slant height of the pyramid and label it l.

Pyramid Dissection

continued

11. Find the total area of the lateral faces. Calculate a value that will be correct for any value of *n*. Write down the calculation and the result.

12. What is this value when the number of faces is at least 50? Why?

Total Area

13. On page "Area," do the necessary constructions, measurements, and calculations to find the total surface area of the regular pyramid, including the base and the lateral faces. Write down your calculation and result.

EXPLORE MORE

14. On page "Net," set *L, R,* and *N* to match the pyramid you made on page "Area." Choose **File | Print Preview** and make sure the net fits on one page. If necessary, click Scale To Fit Page, and then click Print. Label the base and each face with the area you calculated based on the measurements. Cut out the net, fold along the lines, and glue or tape your three-dimensional pyramid together.

15. What does the pyramid look like when the number of lateral faces is large? How could you calculate the surface area of this shape without using the value of *n*? (Your answers to steps 9 and 12 may be useful.)

16. The *height* (*h*) of a pyramid is defined as the vertical distance from the base to the *vertex.* In step 10, you measured the slant height *l* (the height of one of the lateral faces). How can you find the vertical height of the pyramid if you know *l* and *r*? (For instance, if *l* = 15 cm and *r* = 9 cm, what is *h*?) How could you find *l* if you know *h* and *r*? (For instance, if *h* = 5 cm and *r* = 12 cm, what is *l*?)

17. On page "Explore More," you can experiment with the advanced controls that affect the look of this three-dimensional model.

Stack It Up: Volume of Rectangular Prisms

For **GSP5** ACTIVITY NOTES

INTRODUCE

Project the sketch for viewing by the class. Expect to spend about 15 minutes.

1. If centimeter cubes are available, distribute about 25 (at least 20) to each student. Open **Stack It Up.gsp.** Go to page "Layers." **You see one cube.** Drag point L to add one cube at a time until there are five cubes.

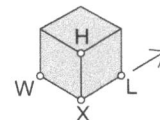

   What is happening? Introduce the language, *row* of cubes, if students don't. For students who need help visualizing the two-dimensional representation as a three-dimensional shape, building this row with centimeter cubes will be helpful.

2. Drag point W so that the row of cubes grows to a layer that has 2 rows of 5 cubes, then 3 rows of 5 cubes, and finally 4 rows of 5 cubes. If students have centimeter cubes, have them represent the growing layer using the cubes. With the addition of each row, ask students what is happening. By focusing on the growth pattern, you'll help students to interpret what they see and to develop a deep understanding of volume.

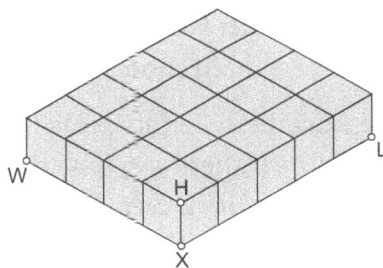

3. *Let's call this one layer. How many cubes fill this layer? How do you know?* Elicit the idea that there are 4 rows of 5 cubes, or $4 \times 5 = 20$ cubes. Because there is only one layer, some students may confuse the answer, 20 cubes, with the measure of the *area* of the top face of the layer. Clarify that 20 cubes tells about the amount of space filled, while area is a measure of the amount of space covered. The area of the top face of the layer is 20 square centimeters. Cut a 4-by-5 rectangle from centimeter graph paper and ask whether this is equal to the space filled by the cubes in the model. [It isn't.]

4. Drag point *H* so that the single layer of cubes grows to 2 layers of 20 cubes each, and then to 3 layers of 20 cubes each. Ask students what is happening.

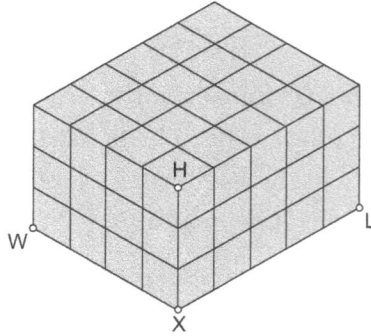

More layers are showing up. They are stacked on top of each other.

The first layer is being copied.

Dragging point H increases the height of the box.

How many cubes fill this box? How do you know? Students may explain that there are 3 layers of 20 cubes, or 3 × 20 = 60 cubes.

5. Distribute the worksheet. Tell students to work through steps 1–6 and do the Explore More if they have time. **In each step, you will use the model to make boxes with a certain number of cubes.** Model enlarging the window so it fills most of the screen. Drag point *X* and let students know they can move a box to make room to build large boxes. **The largest box that can be built using the model is 12 cm long, 12 cm wide, and 12 cm tall.**

6. If students will save their work, model choosing **File | Save As** and tell how they should name their files and where to save them.

DEVELOP

7. Assign students to computers and tell them where to locate **Stack It Up.gsp.** Encourage students to ask a neighbor for help using Sketchpad if needed.

Do not introduce the formula for volume of rectangular prisms (Volume = length × width × height).

Expect students at computers to spend about 30 minutes.

8. Let pairs work at their own pace. As you circulate, observe and listen to students' conversations. Here are some things to notice.

   • In worksheet step 1, students will probably take note of the values for *XL, XW,* and *XH* that appear and update automatically in the sketch. Let students explore these on their own. Some students may begin to make sense of the displayed values and use them; other students may disregard them.

   • In worksheet step 1, some students may not realize that a box with a single row of cubes is one possible solution; they may think that all three dimensions need to be larger than 1 cm.

   • In worksheet steps 1 and 2, challenge students to make a box with no side of length 1 cm. Also notice any students who make the conjecture that the lengths of *XL, XW,* and *XH* must all be factors of the total number of cubes.

The largest box allowed by the model is 12 cm in each dimension. Thus, students cannot build a 1 cm × 1 cm × 24 cm box.

   • In worksheet steps 5 and 6, students are asked to build two different boxes using the same number of cubes in each box. It's fine if students make boxes that share one dimension, such as a 2 cm × 4 cm × 3 cm box and 2 cm × 1 cm × 12 cm box. Prompt those students to extend their thinking by saying, *I wonder if it's possible to make two boxes so that they have no lengths in common.*

   • In the Explore More steps, students are given several new problem types. Encourage students to persevere on their own to try to find solutions. Let students know they will hear others' solutions during the class discussion that follows.

Project the sketch. Expect to spend about 30 minutes.

9. If students will save their work, have them do so now.

SUMMARIZE

10. Gather the class. Students should have their worksheets. *We've been calling our Sketchpad model a box. Mathematicians call it a rectangular prism.* Review the properties of rectangular prisms.

11. Tell the class that the amount of space inside a three-dimensional shape is called the *volume* of the shape. Another way to say this is that volume tells the amount of space a three-dimensional shape takes up. Students have found the volume of boxes by finding the number of cubes that fill the boxes.

Have students practice using the terms *volume* and *cubic centimeter* throughout this discussion.

Go to page "Build a Box" and show one cube. (Students can examine centimeter cubes as well if they have them.) **The unit of measure you have been using is called a** cubic **centimeter. Each edge of this cube is 1 centimeter in length. Each face has an area of one square centimeter.** Drag points to build several small boxes, and ask students to tell the volume of each box expressed in cubic centimeters.

12. Facilitate discussion of methods students have found for determining the volume of rectangular prisms. Begin by saying, I'm thinking of a box that has 2 rows of 10 cubes in the bottom layer, and 5 layers. Write this information on the board.

    How can I figure out the volume of the box? The discussion should bring out the ideas expressed in the student responses that follow.

    First, you have to figure out the number of cubes in a layer. Then multiply that by the number of layers. The box has 2 $\times$ 10 cubes in a layer; that's 20. Times 5 is 100. So the volume of the box is 100 cubic centimeters. This is the same as the way we made boxes with the model. First, you make the bottom layer; then you make more layers that have the same number of cubes.

    To find the number of cubes in a layer, just find the area of the bottom of the box. I think about each cube in the bottom layer sitting <u>on</u> one square centimeter, so the number of square centimeters in the area of the bottom of the box is equal to the number of cubic centimeters in the layer. The area of the bottom of your box is 2 $\times$ 10. Multiply that by the height, 5, which tells you the number of layers.

    Dragging point L creates the length of a box and dragging point W creates the width of a box. We can find the area of the bottom layer by multiplying the length (L) times the width (W). H is the height of the box. After we find the area of the bottom, we multiply by the height.

13. Discuss the Explore More problems, worksheet steps 7–9. In step 9, students may use this reasoning: The largest amount of cubes Roberto can use is 90 (the largest multiple of 5 that is less than 93); 20 layers would require 100 cubes, so 90 cubes will fill two layers of 5 fewer, or 18 layers; and 3 cubes will be left over.

EXTEND

1. *What other questions about building and filling boxes occurred to you? What have you wondered about?* Encourage student inquiry. Here are sample student queries.

   Why don't the large boxes look very realistic?

   Are there patterns that we can use to predict how many different rectangular prisms there are with a certain volume?

   Do boxes with the same volume have the same surface area? If they don't, what's the biggest surface area a box can have for the volume it has?

2. Have students write problems about building or filling boxes. One interesting problem type to introduce if students don't is this: *A box is filled with exactly 30 cubes. The height of the box is 3 cm. What is the length and width of the box?* Students should have no trouble determining that there are 10 cubes in a layer. They may be intrigued by the discovery that it is not possible to determine the box's length and width. Possible dimensions, in centimeters, are 1×10, 10×1, 2×5, and 5×2.

ANSWERS

2. Several answers are possible. Two solutions are 2 layers of 4 cubes, and 1 layer of 3 cubes.

3. Many answers are possible. Two solutions are 3 layers of 4 cubes, and 2 layers of 6 cubes.

4. Because 11 is a prime number, no set of numbers other than 11, 1, and 1 can be used as the dimensions of the box. The two possible solutions are 1 layer of 11 cubes, and 11 layers of 1 cube.

5. Many answers are possible. Two solutions are 6 layers of 4 cubes, and 4 layers of 6 cubes.

6. Many answers are possible. Two solutions are 6 layers of 6 cubes, and 4 layers of 9 cubes.

7. 7 layers

8. This is impossible. No even number is a factor of 27.

9. 13 layers

Stack It Up

Name: _____

Build boxes by making layers of cubes.

EXPLORE

1. Open **Stack It Up.gsp.** Go to page "Build a Box."
   You will drag points *L, W,* and *H* to make boxes.

2. Make a box with exactly 8 cubes. Record how many:
   cubes in each layer _____
   layers _____

3. Make a box with exactly 12 cubes. Record how many:
   cubes in each layer _____
   layers _____

4. Make a box with exactly 11 cubes. Record how many:
   cubes in each layer _____
   layers _____

Go to page "Two Boxes." Now you will build two different boxes, each with the same number of cubes.

5. Make two boxes, each with exactly 24 cubes. Record in the table.

	Cubes in a Layer	Layers	Total Cubes
Box 1			
Box 2			

6. Make two boxes, each with exactly 36 cubes. Record in the table.

	Cubes in a Layer	Layers	Total Cubes
Box 1			
Box 2			

New York City Title I Elementary School Activities Grades 1–5
© 2012 Key Curriculum Press

EXPLORE MORE

Answer these questions. Tell about your thinking. If you want, go back to page "Build a Box" and use the model.

7. Christa used exactly 42 cubes to build a box. Each layer had 6 cubes. How many layers did the box have? _____

8. Marta is making a box with exactly 27 cubes. She wants to use an even number of cubes in each layer. How can she do that?

   _____

   _____

   _____

   _____

9. Roberto has 93 cubes. How many layers high can he build the box if 5 cubes fill each layer?

   _____

   _____

   _____

   _____

Lulu: Introducing the Coordinate Grid

For **GSP5** ACTIVITY NOTES

INTRODUCE

Project the sketch for viewing by the class. Expect to spend about 20 minutes.

1. Open **Lulu.gsp.** Go to page "Trip 1." Enlarge the document window so it fills most of the screen. Distribute the worksheet.

2. Explain, *Today you are going to use Sketchpad to move Lulu, represented by this blue point, to different locations on the grid. Have you seen a grid like this before? If so, where?* Students may reply that they've seen maps or games that have a similar grid, but those grids have letters along one side. *What do those grids help you do?* Elicit the idea that they help you find or name locations.

Depending on the level of your class, you need not formally introduce the terms *x*-axis and *y*-axis at this time.

3. Point out that the Sketchpad grid has a horizontal number line and a vertical number line, which intersect at a location called the origin. *This type of grid is called a coordinate grid. You can use this grid to find and name locations.* Work through worksheet steps 1–4 as a class. Here are some tips.

   - In worksheet step 2, model using the **Arrow** tool to press the up, down, left, and right buttons to move Lulu. *Describe what happens when Lulu moves.* Here are sample student responses.

     Lulu moves sideways or up and down. She doesn't move diagonally.

     When you press each button, Lulu moves in that direction one block.

     Lulu moves up, down, left, or right one space.

     There is a trail of blue dots showing the path she took.

     When you move Lulu, the numbers for her location change.

   - Explain that each "block," or "from one corner to the next corner," is 1 unit. Move Lulu and have students tell how many units and in what direction she has traveled. Repeat this several times, erasing traces after each move.

   - Starting with Lulu at the origin, move her 3 units up. As Lulu moves, have students watch the coordinates. *Two numbers, called coordinates, are used to represent Lulu's location. What happens to Lulu's coordinates when she moves 3 units up?* Students should recognize that the first coordinate stays at 0 and the second coordinate changes to 3. Press *Move Lulu to Origin* and *Erase Traces* to return Lulu to the origin. Now move Lulu 2 units to the right. Again, ask students how Lulu's coordinates change. Offer similar examples

New York City Title I Elementary School Activities Grades 1–5
© 2012 Key Curriculum Press

until students recognize that a horizontal move changes Lulu's first coordinate and a vertical move changes her second coordinate.

- Check for students' understanding. *How do Lulu's coordinates match her location on the grid?* Here are sample responses.

    The first number is the number straight down from Lulu on the horizontal line. The second number is the number straight to the left of Lulu on the vertical line.

    I think of it like a video game. Lulu has to run before she can jump. The horizontal distance is first, and then the vertical distance. For example, if Lulu runs 3 units right and then jumps 2 units up, she'll land at (3, 2).

    The first coordinate tells how many units Lulu is to the right of the start. The second coordinate tells how many units up.

- Be sure students understand that the coordinates tell the distance, horizontally and vertically, from the origin. *Where is the point (0, 0)?* [The point where the horizontal and vertical number lines intersect, the origin.] At this time, also explain that mathematicians have agreed to name the horizontal value first and the vertical value second when writing the coordinates.

- In worksheet step 4, help the class fill out the table. *Let's try to move Lulu to each building. What location do we need to move her to next?* Have volunteers come up and move Lulu to the house, the school, the park, and the store. *Describe the path you took to get Lulu to the building.* Students may change directions several times, moving Lulu first horizontally, then vertically, then horizontally again, for example. Encourage students to try to change directions as few times as possible.

4. *Now you're ready to explore Lulu's trips on the coordinate grid on your own. As you work, think about how the coordinates are affected by how Lulu moves.*

5. If you want students to save their work, demonstrate choosing **File | Save As,** and let them know how to name and where to save their files

DEVELOP

Expect students at computers to spend about 25 minutes.

6. Assign students to computers and tell them where to locate **Lulu.gsp.** Tell students to work through step 19 and do the Explore More if they have time. Encourage students to ask their neighbors for help if they are having difficulty with Sketchpad.

7. Let pairs work at their own pace. As you circulate, here are some things to notice.

   • In worksheet steps 5–9, students explore what happens when Lulu moves 1 unit horizontally and 2 units vertically. Lulu will be able to move only to locations whose y-coordinate is an even number. *What do you notice about the coordinates of the locations Lulu can get to?* If needed, ask students to focus specifically on the second coordinate. Writing down the coordinates as Lulu moves may help some students identify what is happening.

   • Worksheet steps 8 and 9 ask students whether Lulu can reach a point not shown on the grid. This stretches students' thinking; students must reason about Lulu's movements without moving her.

   • In worksheet steps 10–14, students will explore what happens when Lulu moves 3 units horizontally and 1 unit vertically. Lulu will be able to move only to locations whose x-coordinate is a multiple of 3. Note that the coordinates of the school and the park are reversed from each other. You might have students explain the difference between $(4, 6)$ and $(6, 4)$. *Do $(4, 6)$ and $(6, 4)$ name the same point? Explain.*

   • In worksheet steps 15–19, students will explore what happens when Lulu moves 5 units horizontally and 4 units vertically. Lulu will be able to move only to locations that have a multiple of 5 for the x-coordinate and a multiple of 4 for the y-coordinate. Students need to pay attention to both coordinates: Both conditions must be satisfied in order for Lulu to move to a specific location.

By dragging the red point at (1, 0) closer to the origin, students can see more of the grid.

   • In the Explore More, students make their own challenges for classmates to solve. Students move the locations of some or all of the buildings by editing their coordinates (providing additional practice with naming coordinates). Students also decide how far Lulu will move in each direction, and then change the left/right and up/down parameters to correspond. For students who are more familiar with coordinates and don't need practice naming them, you may want to suggest they use page "Make Your Own 2," which allows students to move the buildings quickly simply by dragging them with the **Arrow** tool.

8. If students will save their work, remind them where to save it now.

SUMMARIZE

Project the sketch. Expect to spend about 15 minutes.

9. Gather the class. Students should have their worksheets with them. Go to page "Trip 2." ***Describe how Lulu moves.*** [She moves 1 unit horizontally and 2 units vertically.] ***Can she get to each place? Explain.*** Here are sample student responses.

    Lulu can't get to the park. It's on a point that Lulu skips.

    The park is located at (8, 7). The second coordinate has to be even for Lulu to get there.

    Lulu moves by twos vertically. If I count by twos, I will never hit 7.

    The second coordinate has to be divisible by 2 evenly. The second coordinate for the park is 7, which is not divisible by 2 evenly.

10. Review the movement of Lulu in trips 3 and 4 in a similar manner, if necessary. Ask students questions, making sure they articulate how Lulu moves and why she can or cannot move to certain buildings.

11. If time permits, discuss the Explore More. Have students share challenges they made.

12. ***Why do you need two numbers to describe a location on the coordinate grid? Why does the order in which you write the two numbers make a difference?*** You may wish to have students respond individually in writing to this prompt.

EXTEND

What questions occur to you about Lulu and coordinate grids? Encourage curiosity. Here are some sample student queries.

Can Lulu move to the left of the vertical number line or below the horizontal number line? What would the coordinates be?

How big can the coordinate grid get?

Can Lulu move between units? If she could, what would the coordinates look like?

Why doesn't a coordinate grid use letters along one side like maps do? It would be less confusing.

Can we figure out how far Lulu moves when she travels around?

ANSWERS

3. Lulu moves 1 unit with the press of each button.

4. House: $(3, 2)$, yes; School: $(3, 9)$, yes; Park: $(6, 6)$, yes; Store: $(9, 9)$, yes

6. Lulu moves 1 unit horizontally or 2 units vertically.

7. House: $(3, 2)$, yes; School: $(3, 8)$, yes; Park: $(8, 7)$, no; Store: $(9, 8)$, yes

8. No, she cannot reach $(30, 25)$ because 25 is not an even number.

9. Yes, she can reach $(23, 36)$. She can reach any x-coordinate; she can reach any y-coordinate that is an even number.

11. Lulu moves 3 units horizontally and 1 unit vertically.

12. House: $(3, 2)$, yes; School: $(4, 6)$, no; Park: $(6, 4)$, yes; Store: $(9, 11)$, yes

13. Yes, she can reach $(27, 23)$ because she can reach any x-coordinate that is a multiple of three and any y-coordinate.

14. No, she cannot reach $(31, 21)$ because 31 is not a multiple of 3.

16. Lulu moves 5 units horizontally and 4 units vertically.

17. House: $(5, 4)$, yes; School: $(9, 8)$, no; Park: $(4, 5)$, no; Store: $(10, 12)$, yes

18. No, she cannot reach $(32, 24)$ because 32 is not a multiple of 5.

19. Yes, she can reach $(50, 20)$ because 50 is a multiple of 5 and 20 is a multiple of 4.

21. Answers will vary.

Lulu's Trips

For GSP5   Name: _____

Help Lulu travel around town.

EXPLORE

1. Open **Lulu.gsp.** Go to page "Trip 1."

2. Practice moving Lulu around the grid. Press the buttons to move Lulu left, right, up, and down.

3. How many units does Lulu move in each direction on one button press?

   _____

   _____

4. Try to move Lulu to each place. Use the table to record her trips.

   Trip 1

Place	Location	Can Lulu Reach It? (Yes or No)
House		
School		
Park		
Store		

5. Go to page "Trip 2." Practice moving Lulu around the grid.

6. How many units does Lulu move in each direction on one button press?

   _____

   _____

7. Try to move Lulu to each place. Use the table to record her trips.

Trip 2

Place	Location	Can Lulu Reach It? (Yes or No)
House		
School		
Park		
Store		

8. Can Lulu reach (30, 25)? Why or why not?

_____

_____

9. Can Lulu reach (23, 36)? Why or why not?

_____

_____

10. Go to page "Trip 3." Practice moving Lulu around the grid.

11. How many units does Lulu move in each direction on one button press?

_____

_____

New York City Title I Elementary School Activities Grades 1–5
© 2012 Key Curriculum Press

12. Try to move Lulu to each place. Use the table to record her trips.

Trip 3

Place	Location	Can Lulu Reach It? (Yes or No)
House		
School		
Park		
Store		

13. Can Lulu reach (27, 23)? Why or why not?

_____

_____

14. Can Lulu reach (31, 21)? Why or why not?

_____

_____

15. Go to page "Trip 4." Practice moving Lulu around the grid.

16. How many units does Lulu move in each direction on one button press?

_____

_____

17. Try to move Lulu to each building. Use the table to record her trips.

Trip 4

Place	Location	Can Lulu Reach It? (Yes or No)
House		
School		
Park		
Store		

18. Can Lulu reach (32, 24)? Why or why not?

_____

_____

19. Can Lulu reach (50, 20)? Why or why not?

_____

_____

EXPLORE MORE

20. Go to page "Make Your Own 1." Make a problem for a classmate to solve.

Change how far Lulu moves in each direction on one button press: Type a new number for *left/right* or *up/down*.

Change the location of a building by pressing its *Edit* button and entering new coordinates. When you're finished, press *Done*.

For
GSP5

21. Describe the problem you made. How far does Lulu move in each direction on one button press? Can she reach the buildings? Why or why not?

_____

_____

_____

_____

_____

_____

_____

Coordinate Patterns: Points on a Line

For GSP5 ACTIVITY NOTES

INTRODUCE

Project the sketch on a large-screen display for viewing by the class. Expect to spend about 10 minutes.

1. Open **Coordinate Patterns.gsp.** Go to page "Patterns."

   Introduce the model by pressing the + and − buttons. After each press, ask students to describe how the point has moved. Here are two sample descriptions: *The point moved one to the right. The point moved down one.*

   Ask questions such as these to facilitate a review of identifying points by ordered pairs of numbers.

   Where is the point now?

   What does it mean to say that its coordinates are $(3, 4)$*?*

   What is the x-coordinate? What is the y-coordinate?

   What do you predict will happen to the point and its coordinates when I press this + button?

2. *Today you're going to move the point to locations on the grid that fit certain rules. Your challenge will be to describe and make sense of the results. We'll do one rule together; the rest will be up to you.* Distribute the worksheet and direct students' attention to the first rule in step 2: The x- and y-coordinates are equal. *Our job is to move the point to places that fit this rule.*

If students describe a less systematic approach, try it. Students will develop more systematic ideas as they proceed through the worksheet.

3. Ask the class to name some locations on the grid that fit the rule. Record the suggestions. Possibilities include $(0, 0)$, $(1, 1)$, $(2, 2)$, and $(3, 3)$.

   Ask the class how to move the point to the locations they named. One method is to move the point from $(0, 0)$ to $(1, 1)$ by moving 1 unit over, or right (pressing the + button corresponding to the x-coordinate) and then moving 1 unit up (pressing the + button corresponding to the y-coordinate). Following this same "over 1 and up 1" approach lands the point at $(2, 2)$, $(3, 3)$, $(4, 4)$, and so on.

   Whenever the point lands where the x- and y-coordinates are equal, press the *Mark Me* button. Doing so leaves behind a trace of the point. In this way, all points on screen where the x- and y-coordinates are equal can be viewed at once. Note that $(0, 0)$, where the point began, is also a place to mark.

4. With at least four points marked, ask, *What patterns do you see?* One pattern is that the points all seem to be on a line. Press *Show Graph a.* A line appears that includes every point whose x- and y-coordinates are equal.

340

New York City Title I Elementary School Activities Grades 1–5
© 2012 Key Curriculum Press

5. Highlight these Sketchpad tips by modeling them.

   *Press **Reset** to start over if you want to.*

   Only mark points at locations that now show on screen.

   *Don't drag the numbers along the axes. The result, because of the programming, will not be useful. If you forget and drag, choose **Edit | Undo**.*

   Don't scroll to view more of the grid. This will cause existing traces to disappear. (Traces also disappear when you move to another page of the sketch.)

DEVELOP

Expect students at computers to spend about 30 minutes.

6. Assign students to computers and tell them where to locate **Coordinate Patterns.gsp**. Explain that students should first experiment with the model in order to become comfortable with the + and − buttons and their effects on the point's movement and location. Students should then work through step 12 of the worksheet.

7. As you circulate, observe students' approaches and ask questions to learn about their thinking.

In rule 2b, for example, students might mark their locations in the order (4, 8), (1, 2), (5, 10).

   • Allow students to choose the order of their moves: horizontally and then vertically, or vertically and then horizontally. You may find that students intuitively vary the order depending upon the rule they are working with and how they are making sense of it. Clarify, as needed, that ordered pairs give the *x*-coordinate first, but students can move in either order on the grid.

   • Students are likely to use a variety of approaches to find points that satisfy the rules. Some students may tend to use the relationships of the numbers in the ordered pairs; some may tend to use the movement patterns; and some may look for the next place that will extend the implied line, realizing that once they have two points, they know where the line will be.

   • As students begin to mark points that fit a rule, some students may move the point in a meandering fashion around the grid. After students have had an opportunity to work in this way, ask, ***Can you find an orderly and efficient way to locate and mark points that satisfy the rule?*** Let students do the thinking about what "orderly" and "efficient" mean in this context. One aspect of being efficient is having a way to move from a location to a nearby location without needing to double back later to mark a location that was missed.

- As students explore button presses that move the point from one location to the next, they will discover systematic ways to travel. In rule 2b, for example, students starting at $(0, 0)$ can move the point "right 1 and up 2" to arrive at the next nearest point, $(1, 2)$. Repeating the same "right 1 and up 2" process moves the point to $(2, 4)$, $(3, 6)$, $(4, 8)$, and so on—all locations where the y-coordinate is twice the x-coordinate.

- Once students realize that such systematic movement patterns exist, the nature of exploration will likely change for them. Knowing, for example, to move "right 1 and up 2" for rule 2b frees students from checking the coordinates of each new point to see whether the y-coordinate is twice the x-coordinate: The pattern of button presses guarantees that it will be.

These patterns students are exploring relate directly to the slopes of the lines on which the points lie.

- When students have worked with several rules in worksheet step 2, they will notice that the rules have something in common: The plotted points that satisfy the rule lie on a line. Students may notice that all the lines pass through the origin and that each line has a different steepness. Ask why this is so. ***Can you think of a reason why the points all lie on a line? Why does the line pass through the origin?***

- You may observe some students leaving "gaps" in their marked points and then checking that the line goes through these places that they could have marked but did not. ***Please explain to me your approach.*** Students who use this approach are likely to have focused on the numerical relationships and understood that the skipped points could be interpolated. They may also show that they understood the pattern of movement in this method (for example, that "over 1, up 1" will eventually get you to the place that can be reached, with gaps, by traveling "over 3, up 3").

If students drag the grid, have them choose **Edit | Undo** to back up until the grid returns to its original appearance.

- Steps 5 and 9 ask students to create their own rules similar to the ones in steps 2 and 7. Depending on the rules students devise, they may or may not be able to mark locations that satisfy them. The rule that the x-coordinate is 40 times the y-coordinate, for example, satisfies locations like $(40, 1)$ and $(80, 2)$, which do not appear on the limited view of the coordinate grid shown on the screen.

> • Notice how students' conversations with each other and with you are promoting their use and understanding of math vocabulary for terms related to coordinate geometry.

Explore More

8. If time is running out and some students have not reached worksheet step 12, ask them to skip ahead. This step repeats the rules used in worksheet step 7, but now students work in the third quadrant of the grid, where the *x*- and *y*-coordinates are negative. Students will need to think carefully about what it means for the *x*-coordinate of a point to be more or less than the *y*-coordinate. Ultimately, students are forming the same lines that they did in step 7, but now they are viewing the extension of those lines into another quadrant of the grid.

SUMMARIZE

Project the sketch on a large-screen display for viewing by the class. Expect to spend about 20 minutes.

9. Gather the class. Ask volunteers to describe strategies they used to find points that satisfied each rule as well as the patterns they noticed. Engage the class in discussion of each strategy, making it the students' responsibility to verify the identified strategy and to ask for and contribute to clarification of the explanations.

   Be sure to discuss the efficient + and − patterns that allowed students to move from one location to the next. In all cases, a repeated pattern of "right and up" or "up and right" movements generated a line. (For your own information, this is a key observation for students when they study the slopes of lines more formally.)

10. If students haven't posed questions like the two that follow, introduce them yourself.

    Why does moving "up 2 and right 1" (or "right 1 and up 2") guarantee that the y-coordinate of the new point will be twice its x-coordinate? Provide lots of time for students to work on communicating their reasoning to the class.

    The class might also consider the following situation. Suppose two people (choose two students and use their names) each have the same number of marbles. They can match their marbles one-for-one. Every time Jane puts out one more marble and Juan puts out one more marble, each will have the same number of marbles out. Now suppose instead that, to start, Jane has twice as many marbles as Juan. For every one marble Juan has, Jane can put two alongside it. Juan increases the number of marbles he has out by one and Jane increases hers by two. They can continue to do a one-to-two pairing.

Is it true that the y-coordinate of the new point will be twice its x-coordinate no matter where you start? This is true when the starting location is one in which the *y*-coordinate is twice the *x*-coordinate. For Jane and Juan, it will be true that Jane has twice as many marbles out as Juan if they started with a two-to-one pairing.

- When discussing worksheet step 8, if students don't introduce the idea, show all of the lines on pages "Patterns" and "Patterns 2." Give students time to study the two pages as you toggle between them. Ask for observations. Students may note that all of the lines for the first set of rules go through the origin, whereas only one line from the second set of rules does that. (Students may also observe that the lines on page "Patterns 2" all have the same "steepness." Invite discussion of these ideas.

- Informally assess students' development of math vocabulary. Are more students secure in their use and understanding of the terms?

Explore More

11. Worksheet step 12 provides an opportunity for students to extend their understanding of negative numbers. Elicit the idea that students have formed the same lines that they did in step 7, but now they are viewing the extension of those lines into another quadrant of the grid. How do they make sense of that? What does it mean for the *x*-coordinate of a point to be more or less than the *y*-coordinate?

EXTEND

1. Have students consider whether there are points other than those with integer coordinates that sit on the lines. ***Are there places on the line you can't reach no matter which + or − buttons you press? How many places are there?*** Students may note that there are infinitely many points that sit on their lines whose coordinates are not integers. In the case of the line whose *x*- and *y*-coordinates are equal, for example, these points include $\left(\frac{1}{3}, \frac{1}{3}\right)$, $(0.2, 0.2)$, $(-1.5, -1.5)$ and (π, π).

2. Invite students to share questions that occurred to them while they were working, or to pose some now. ***What other questions can you think of, even if you can't answer them now?*** Invite students to share their first thoughts about any questions posed, and then plan to provide additional time later for students to explore questions that interest them. Here are some sample student questions.

Could a point satisfy more than one rule?

Could we mix directions, moving, for example, "right, up, right"?

How many ways are there to get from one point to the next without going backward? What does it mean to "go backward"?

Which points are only, say, three button presses apart?

Are there rules whose points don't lie on lines? What would those rules look like? What shapes would the points form?

ANSWERS

3. a. Possible locations include $(0, 0)$, $(1, 1)$, $(2, 2)$, $(3, 3)$, and $(4, 4)$.
   b. Possible locations include $(0, 0)$, $(1, 2)$, $(2, 4)$, $(3, 6)$, and $(4, 8)$.
   c. Possible locations include $(0, 0)$, $(1, 3)$, $(2, 6)$, $(3, 9)$, and $(4, 12)$.
   d. Possible locations include $(0, 0)$, $(2, 1)$, $(4, 2)$, $(6, 3)$, and $(8, 4)$.
   e. Possible locations include $(0, 0)$, $(3, 1)$, $(6, 2)$, $(9, 3)$, and $(12, 4)$.

4. Students should notice that the plotted points that satisfy the rule lie on a line. They may notice that all the lines pass through the origin, and that each line has a different steepness.

5. Students' rules will vary.

6. Ordered pairs that fit students' rules will vary.

7. a. Possible locations include $(0, 1)$, $(1, 2)$, $(2, 3)$, and $(3, 4)$.
   b. Possible locations include $(0, 2)$, $(1, 3)$, $(2, 4)$, and $(3, 5)$.
   c. Possible locations include $(0, 3)$, $(1, 4)$, $(2, 5)$, and $(3, 6)$.
   d. Possible locations include $(1, 0)$, $(2, 1)$, $(3, 2)$, and $(4, 3)$.
   e. Possible locations include $(2, 0)$, $(3, 1)$, $(4, 2)$, and $(5, 3)$.
   f. Possible locations include $(3, 0)$, $(4, 1)$, $(5, 2)$, and $(6, 3)$.

8. As with the rules in step 2, each rule in step 7 describes points that lie on a line. The lines in step 7, however, do not pass through (0, 0). All of the lines in step 7 have the same "right 1 and up 1" (or "up 1 and right 1") pattern for moving from one location to the next.

9. Students' rules will vary.

10. Ordered pairs that fit students' rules will vary.

11. Students will likely notice several patterns. One pattern focuses on the locations of the point that satisfy the rules. These locations all sit on lines. Another pattern focuses on the consistent $+$ and $-$ button presses that move the point from one location to the next. As described in these notes, for each rule there is a systematic pattern of button presses that allows students to pilot from one location to the next. In step 7, the pattern was always the same: "right 1 and up 1" (or "up 1 and right 1").

12. a. Possible locations include $(-1, 0)$, $(-2, -1)$, $(-3, -2)$, and $(-4, -3)$.
    b. Possible locations include $(-2, 0)$, $(-3, -1)$, $(-4, -2)$, and $(-5, -3)$.
    c. Possible locations include $(-3, 0)$, $(-4, -1)$, $(-5, -2)$, and $(-6, -3)$.
    d. Possible locations include $(0, -1)$, $(-1, -2)$, $(-2, -3)$, and $(-3, -4)$.
    e. Possible locations include $(0, -2)$, $(-1, -3)$, $(-2, -4)$, and $(-3, -5)$.
    f. Possible locations include $(0, -3)$, $(-1, -4)$, $(-2, -5)$, and $(-3, -6)$.

Coordinate Patterns

For GSP5 Name:

Move a point to locations that fit a given rule. Look for patterns.

EXPLORE

1. Open **Coordinate Patterns.gsp.** Go to page "Patterns."

2. Choose one of these rules.
   a. The *x*- and *y*-coordinates are equal.
   b. The *y*-coordinate is twice the *x*-coordinate.
   c. The *y*-coordinate is three times the *x*-coordinate.
   d. The *x*-coordinate is twice the *y*-coordinate.
   e. The *x*-coordinate is three times the *y*-coordinate.

   Now you will place some points that fit the rule.

   Press the + and − buttons to move the red point.

   Find and mark at least four points that fit the rule. (Press *Mark Me* to leave a trace of the point behind.)

   Record the *x*- and *y*-coordinates of the marked points in the tables below.

   Press *Show Graph* for the rule. Note what you see.

   Press *Reset.*

3. Repeat step 2 for the other rules. Record the marked points.

Rule 2a		Rule 2b		Rule 2c		Rule 2d		Rule 2e	
x	*y*	*x*	*y*	*x*	*y*	*x*	*y*	*x*	*y*

4. What do you notice about the points for each rule? (*Hint*: Press all of the *Show Graph* buttons.)

_____

_____

5. Write a new rule like the ones in step 2.

_____

_____

6. Mark on the grid at least four points that fit your rule. Record the points using ordered pairs.

_____

_____

7. Go to page "Patterns 2." Mark at least four points that fit each of these rules. Record the marked points in the tables.
   a. The *y*-coordinate is one more than the *x*-coordinate.
   b. The *y*-coordinate is two more than the *x*-coordinate.
   c. The *y*-coordinate is three more than the *x*-coordinate.
   d. The *x*-coordinate is one more than the *y*-coordinate.
   e. The *x*-coordinate is two more than the *y*-coordinate.
   f. The *x*-coordinate is three more than the *y*-coordinate.

Rule 7a		Rule 7b		Rule 7c		Rule 7d		Rule 7e		Rule 7f	
x	*y*	*x*	*y*	*x*	*y*	*x*	*y*	*x*	*y*	*x*	*y*

8. Compare the rules in step 7 to the rules in step 2. How are they the same? How are they different?

_____

_____

9. Write a new rule like the ones in step 7.

_____

_____

10. Mark on the grid at least four points that fit this new rule. Record the points using ordered pairs.

_____

11. What kinds of patterns did you find as you worked with the rules in steps 2 and 7?

_____

_____

_____

_____

_____

_____

_____

Coordinate Patterns

continued

EXPLORE MORE

12. Go to page "Explore More." You are looking at a different part of the grid.

    Move the point to fit these rules. Mark the points.

    Don't leave this part of the grid.

    Record your work and tell about what you discover.

    a. The y-coordinate is one more than the x-coordinate.
    b. The y-coordinate is two more than the x-coordinate.
    c. The y-coordinate is three more than the x-coordinate.
    d. The x-coordinate is one more than the y-coordinate.
    e. The x-coordinate is two more than the y-coordinate.
    f. The x-coordinate is three more than the y-coordinate.

Finish the Polygon: Concept of Area

INTRODUCE

Project the sketch. Expect to spend about 10 minutes.

1. Open **Finish the Polygon Present.gsp.** Go to page "Example 1." Ask students to describe what they see. Here are sample student responses.

   The shape is a rectangle.

   The shape is a polygon.

   The sides of the rectangle sit on the gray lines.

   Two sides of the rectangle have a length of 3. The other two sides have a length of 5. I can tell by counting the squares.

2. Discuss how the squares on the grid can be used to measure the polygon. Each side of a square has a length of 1 unit, so the dimensions of the polygon are 3 units by 5 units. Ask, **What is the area of the polygon?** Some students may find the area by counting the squares inside the rectangle. Others may multiply 3 by 5 to get 15. Because squares are being used as units to measure area, the area of the polygon is 15 square units.

3. Go to page "Example 2." Now the polygon shown is not a rectangle. Its sides are not all the same length, nor are its angles all the same size. (Students may not realize at first that not all of the angles are right angles.) **What is the area of this polygon?** Students who have thought about area only in the context of rectangles may wonder whether this polygon even has an area. This is an opportunity to facilitate an exchange of ideas among students and let some students make the case that area is the amount of flat space a shape covers. Counting squares is a useful way to find the area of the rectangle, and that method works here too. Because there are 11 squares in the polygon's interior, its area is 11 square units.

4. Still on page "Example 2," use another part of the grid to demonstrate how to draw a polygon using the **Segment** tool. Tell students you want to make a polygon with an area of 7 square units and "think out loud" as you are deciding how you will do that.

5. Demonstrate ways to change a drawing.

   • To make the segments thick, choose **Display | Line Style | Thick.** To make the segments black, choose **Display | Color.**

   • To undo actions step-by-step, choose **Edit | Undo** one or more times.

- To make a segment longer or shorter, drag one of its endpoints.

- To delete a segment, select the segment and choose **Edit | Clear Segment** or press the Delete key on your keyboard.

6. Go to page "Polygon 1" and read the directions with the class. Demonstrate adding several sides to the polygon by drawing segments.

7. If you want students to save their work in the document, demonstrate choosing **File | Save As,** and let them know how to name and where to save their files.

DEVELOP

Expect students at computers to spend about 35 minutes.

8. Assign students to computers and tell them where to locate **Finish the Polygon.gsp.** Distribute the worksheet. Students should follow steps 1–4 and do the Explore More if they have time. Encourage students to ask a neighbor for help if they have questions about using Sketchpad.

9. As you circulate, observe how students go about making their polygons and finding the area. Here are some things to note.

- How are students finding the area of polygons? Are they counting unit squares one by one? Are they dividing a polygon into rectangular areas, finding the area of each rectangle, and adding the areas?

- Do students recognize that there are many ways to complete each polygon so that it has an area of 13 square units?

- For students who find worksheets steps 3 and 4 difficult, suggest they begin by drawing *any* polygon that passes through the point. Then, after finding the polygon's area, students can change the polygon by adding, removing, or relocating segments.

- On page "Polygon 4," when students have constructed a polygon with area 13 square units, ask, ***How can you change your polygon so the area is 12 square units and point A is still on the border?*** Invite students who want more challenge to pose other questions of this type and try to answer them as they work on pages "Polygon 4," "Polygon 5," and "Polygon 6."

10. Have students save their work so that it can be shared with the class. Collecting sketches on a flash drive makes it easy to display them from the shared computer in the class discussion that follows. Viewing more

than one polygon that solves each problem helps students develop a
stronger conceptual understanding of area.

SUMMARIZE

Project the sketch. Expect
to spend about 30 minutes.

11. Gather the class. Discuss several worksheet problems, viewing two or
more ways to complete each polygon. Share saved sketches, or have
students come to the computer and quickly create polygons.

12. ***Today you explored the area of some polygons that are not rectangles.
What have you learned about area?*** Here are sample student responses.

A polygon that is not a rectangle has area.

*When a polygon is not a rectangle, you can't multiply the number of rows
times the number of columns to find the area.*

*When a polygon is not a rectangle, you can't multiply the length times the
width to find the area.*

*One way to find the area of polygons like the ones we made is to count the
square units inside the polygon.*

*Another way to find the area is to divide the polygon into rectangular areas,
find the area of each of those, and then add all the areas together.*

EXTEND

1. ***What questions occurred to you about finding area?*** Encourage
curiosity. Here are some student queries.

How many polygons are there with an area of 13 square units?

*On page "Polygon 5," can you make a polygon through point A that has an
area of less than 13 square units?*

Can you tell the area of a polygon by counting the number of sides?

How can you find the area of shapes that have curved sides?

2. Have students create problems like those in worksheet steps 3 and 4.
Students may enjoy figuring out how far they can place point A from
the partially complete polygon and have a solution be possible.

3. Have students open a new copy of **Finish the Polygon.gsp** and work through the worksheet again with two changes to the rules:

    • Sides do not have to lie along the grid lines; they may be drawn diagonally from one intersection to another on the grid.

    • Polygons must be coverable by whole or half square units; if they aren't, students must be able to justify the areas for the polygons they construct. This illustration shows student work in which the student has chosen the **Text** tool and double-clicked in each part of the polygon to make a caption with the area of the part.

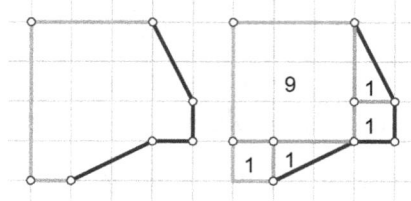

ANSWERS

Answers will vary. Some possible answers are shown here.

1.

2.

3.

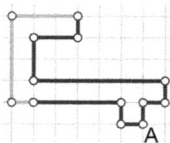

4.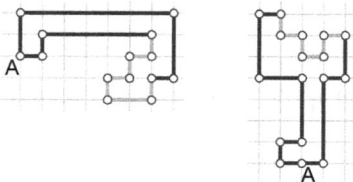

Finish the Polygon

Explore the area of some polygons that are not rectangles.

CONSTRUCT

1. Open **Finish the Polygon.gsp.** Go to page "Polygon 1."
   A drawing of a polygon has been started.
   Finish the polygon in any way you like. There are two rules:
   • The area of the polygon must be 13 square units.
   • All sides of the polygon must be on the grid lines.
   To change your polygon, you can
   • Drag the endpoint of a segment to a new place.
   • Choose **Edit| Undo** to back up.
   • Select a segment and choose **Edit| Clear Segment.**

2. Go to pages "Polygon 2" and "Polygon 3."
   Follow the two rules and finish each polygon.

3. Go to page "Polygon 4." Again, follow the two rules.
   Challenge: Point A must lie on a side or at a vertex of the polygon.

4. Go to pages "Polygon 5" and "Polygon 6" and do the same.

EXPLORE MORE

5. Go to page "Make Your Own."
   Start a polygon for a partner to finish.

Angle Measurement: Estimation Practice

INTRODUCE

Project the sketch on a large-screen display for viewing by the class. Expect to spend about 10 minutes.

1. Open **Angle Estimation.gsp.** Go to page "The Model." *In this activity you'll use a Sketchpad model to sharpen your skills at estimating the size of angles.* Explain that this page introduces the model. Invite a volunteer to follow the directions on the page as the class observes.

2. Go to page "How to Play 1." Explain that this page and the next page tell students how to use the model. When they work on their own, students can refer to these pages to be reminded of what to do.

   Invite a volunteer to follow the directions on this page and on page "How to Play 2" as the class observes.

3. Distribute the worksheet and explain how students will use it to record the work they do independently at computers.

DEVELOP

Expect students at computers to spend 15 to 30 minutes.

4. Make sure students know where to locate **Angle Estimation.gsp** when they work independently.

5. As students work, you may want to observe how they approach the estimation challenges.

   • Do they think about benchmark angles?

   • Do they tend to orient the angles with one side horizontal or vertical, or are they comfortable with different orientations?

   • Do they estimate reflex angles by estimating the size of the smaller angle and subtracting that number from 360°?

   • Are they refining and improving their estimates as they receive feedback from the computer?

SUMMARIZE

Project the sketch on a large-screen display for viewing by a small group. Expect to spend 10 to 15 minutes.

6. After some students have used the sketch, provide an opportunity for them to discuss the strategies they used and to assess whether they improved in their ability to estimate angle measurements.

How Close Are You?

For GSP5 Name:

Sharpen your skills at estimating angle sizes.

1. Open **Angle Estimation.gsp.**

2. Go to page "Estimate 1." For each angle you make, record your estimate in the first table. Then record the actual measurement.

3. Go to page "Estimate 2." In the second table, record the measurement you are trying to show. Then record the actual measurement of the angle you make.

Your Estimate	Actual Measurement	Measurement to Estimate	Actual Measurement

Point Graphs: Representing Data

INTRODUCE

Project the sketch for viewing by the class. Expect to spend about 30 minutes.

1. Present the following situation. *A class is planning to make popcorn balls for sports day at their school. The class that made the popcorn balls last year says that one cup of raw (uncooked) popcorn will make four popcorn balls.*

Use a Data Table

2. *The students in the club started this table. Let's complete it.* Show the recording of this partially filled table, which you have prepared on chart paper or a transparency. Have students explain their reasoning as the class completes the table.

Raw Popcorn (in cups)	Popcorn Balls
1	4
2	
4	
8	

3. *What patterns do you see in the table?* The columns in the data table show that the number of popcorn balls doubles as the number of cups of raw popcorn doubles. The rows in the data table show that the number of popcorn balls is always four times the number of cups of raw popcorn.

Choose a Type of Graph

4. Tell students that some of the club members wonder whether making a graph might help them answer questions that will come up. For example, they know that last year 102 popcorn balls were sold. How many cups of raw popcorn, they wonder, would be needed to make that many popcorn balls? *What would be the best kind of graph for the club to use?* The discussion should bring out the ideas that a point graph is often used when both variables are numbers and when you want to know about values between the data points.

5. Open **Point Graphs.gsp** and go to page "Popcorn." Explain that the class will learn to use Sketchpad to create a point graph.

If the class can justify
the use of the scales
on a different page
of the sketch, use
that page rather than
page "Popcorn."

Label the Axes

6. Discuss which variable should be graphed on each axis. We'll assume here that the class graphs the values from the first column of the data table on the horizontal axis, and the second column on the vertical axis. Refrain from labeling the axes until step 8 below.

7. Discuss the scales of the axes. Students should note that the horizontal and vertical axes are scaled differently. ***Will the numbers on the axes work well for what we want to graph?*** Go to several other pages and discuss the numbers on the axes. ***Would these be more appropriate?*** Elicit students' thinking about the range of the data, the appropriateness of the intervals shown on the axes, and room for *extrapolation* (if students are familiar with making predictions for values that go beyond the existing data).

   Students may suggest that they will be able to scroll to see more of the graph if necessary. Confirm that when they have created their point graphs and a line through the data, they will be able to scroll in their sketches. Note that you can adjust the scale of an axis by dragging the tick labels. This is another way to see more of the graph on screen.

It is not possible to rotate text 90° to label the *y*-axis.

8. When the class has decided on the scales to use, use the **Text** tool to label the axes by creating a caption with the name of the variable for each axis. (You may want to add a letter to stand for the variable.) Ask the class to suggest a title for the graph. Create a caption containing the title near the top of the screen.

Plot Points

9. Model graphing the data in the data table. Choose **Graph | Plot Points.** In the Plot Points dialog box that appears, enter the coordinates for the first point you want to plot, (1, 4), and click **Plot.** Repeat to plot the remaining data in the table. Discuss including (0, 0) as a data point and have students observe that there is already a point at the origin of the Sketchpad grid. When all points are plotted, click **Done.**

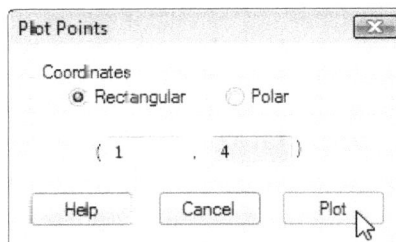

10. Using the **Arrow** tool, select all of the data points. Choose **Display | Color** and choose a color other than red. (This will help students distinguish the data points from other points in the graph.) Make sure students can see the points in the new color.

Look for Patterns

11. ***What do you notice about these points?*** Students should observe that each point is "over and up" the same amount from the previous point, that the points go up and to the right, and that the points lie along a line.

    Let's fit a line to the points. Using the **Line** tool, click in empty space on the grid to locate a first point through which the line passes. Click again in empty space to locate a second point through which the line travels.

    Invite a volunteer to adjust the line by dragging the red points. Let students confirm that the line goes through all the data points; the points lie exactly on a straight line. Point out that the point at (0, 0) also lies on the line. ***Does that make sense? Why or why not?*** A sample student explanation is this: *If no popcorn is used, then no popcorn balls are made.*

Predict

12. Pose this question: ***One of the club members said, "I want to know how much popcorn to use to make 10 popcorn balls."*** Have students suggest how to use the graph to answer the question.

13. Demonstrate how to interpolate using the dashed red lines at the left and bottom edges of the screen.

    - Drag the horizontal red line up until you reach 10 on the vertical axis (number of popcorn balls). Observe that the red line extends through the line through the data points.

    - Drag the vertical red line to the right until it meets the line at the point where the red horizontal line also meets it. Observe that the vertical red line extends through the horizontal axis (number of cups of popcorn).

    - With the class, estimate the value of the location where the red line meets the horizontal axis. [The location is exactly 2 1/2.]

Go back to the data table and ask students whether using 2 1/2 cups of popcorn for 10 popcorn balls makes sense. Here are samples of student reasoning.

It makes sense because 2 1/2 is halfway between 2 and 3, and 10 is halfway between 8 and 12.

From 8 popcorn balls, it takes a cup to make 4 more, so it only takes half a cup to make 2 more.

14. ***Can we figure out how many cups of popcorn would be needed to make 20 popcorn balls?*** Have students talk in pairs or small groups, and then as a class. Discussion should bring out that because 20 balls is twice 10 balls, twice 2 1/2 cups, or 5 cups, would be needed.

    Let's see whether we get the same answer using the graph. How can we use the red lines to check? Invite a volunteer to help you at the computer. The lines may be dragged in either order. For the sake of example, let's assume here that the vertical red line is dragged to 5 on the horizontal axis (number of cups of popcorn); now the horizontal red line can be dragged up to meet the vertical red line at the line through the data points. ***Where does the horizontal line cross the vertical axis (number of popcorn balls)?*** [At 20]

15. Invite the class to propose questions that can be answered using extrapolation. Choose a question and invite a volunteer to the computer to drag the red lines in order to use the graph to answer the question. The class should suggest scrolling to the right and/or up, if needed.

16. Model using **Edit | Save As** to rename the file and save it. This is a convenient stopping place if you want to do this activity over two days.

DEVELOP

Continue to project the sketch. Expect to spend about 20 minutes.

17. Go to page "Cookie Sales." ***This graph shows the money a class made when they sold cookies over eight days for a fundraiser.*** Ask students to describe the graph. ***What do you notice about these data points?*** Students should observe that the points are "around a line" and go in an uphill direction.

Create a Line that Approximates the Data

18. Construct a line in the grid as you did in the previous graph. Model dragging the two points on the line to move the line.

19. Have a volunteer come to the computer to position the line so that it appears to fit the data points as closely as possible. If students have other ideas about where the line should be positioned, invite them to the computer to position the line.

Predict

20. Have the class pose questions that can be solved by extrapolating. Invite volunteers to drag the red lines to make predictions in response to several of the class's questions.

SUMMARIZE

Project the sketch. Expect to spend about 10 minutes.

21. Distribute the worksheet so students can preview it for use at another time. Tell them that they will have a copy of the worksheet when they use Sketchpad in a graphing activity in the future. Give students time to read through the steps while today's demonstration is fresh in their minds.

22. Facilitate an exchange of ideas about the use of technology to create and analyze data. If students have used other software for data investigations, include discussion of those technologies as well.

EXTEND

1. Open **Point Graphs Present.gsp.** Go to page "Extend" to explore another way to interpolate and extrapolate. Drag point *A* and notice that its coordinates update. Use the displayed coordinates to find data points between and beyond the initial data points. You will want to discuss rounding with students when they see that Sketchpad, because of the way it's built, will sometimes display the same coordinates for two points that are near each other.

If students will use this method in another graph, give these directions.

- Using the **Point** tool, construct a point *A* on the line you have drawn through the data points (or, as appropriate, the line that fits the data points as closely as possible).

- With the point selected, choose **Measure | Coordinates.**

- Drag the point along the line. Use the displayed coordinates to find data points between and beyond the initial data points.

2. When using Sketchpad to create point graphs for data investigations, here are some additional options available to students.

- If students will plot data points with values other than whole numbers, have them choose **Edit | Preferences** and, in the Units panel, set the precision for **Others** to **tenths** or **hundredths,** as appropriate.

- Have students scale the axes themselves by dragging the unit point or tick labels on each axis.

- Have students graph additional data sets on a graph following the directions in worksheet step 9.

- When graphs are complete, have students prepare them for display. If students will print their graphs, make sure they choose **File | Print Preview** and, in the dialog box that appears, set the image to fit on one page and then click **Print.**

Point Graphs

Use Sketchpad to make a point graph.

LABEL THE AXES

1. You should have a table of data. Decide which variable goes on each axis of the graph you will create.

2. Open **Point Graphs.gsp.**

   Look through the pages to see the ways the axes are numbered. Choose a page for graphing your data.

3. Create captions to label each axis and title your graph.

PLOT POINTS

4. Now you will plot your data.

   Choose **Graph| Plot Points.**

   In the dialog box that appears, enter the coordinates for the first point you want to plot, and click **Plot.**

   Repeat to plot the rest of the data.

   When all points are plotted, click **Done.**

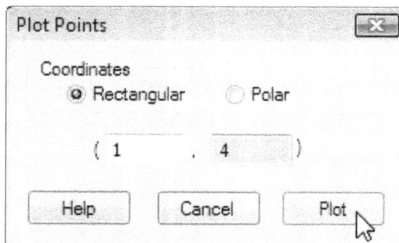

```
Plot Points                                [X]

 Coordinates
   ⦿ Rectangular        ○ Polar

        (  1     .    4      )

   [ Help ]    [ Cancel ]    [ Plot ]
```

LOOK FOR PATTERNS

5. Do the data points lie on a line? If they do, draw a line going through all the points.

   Does the point at (0, 0) lie on the line? Decide whether it makes sense to include that point as a data point.

New York City Title I Elementary School Activities Grades 1–5

6. If the data points do not lie on a line, do they lie close to a line? If so, draw a line and then, using the **Arrow** tool, drag the points on the line so that the line fits the data points as closely as possible.

PREDICT

7. Use the dashed red lines to find the locations of data points other than those you plotted.

8. Scroll up and/or to the right if you need to see more of the graph.

OTHER DATA SETS

9. Do you have another set of data to plot on the same graph? If so, follow steps 3–6.

    For each set of data you plot, use a new color for the data points and for the graphed line or line of best fit.

    Create a caption to label each set of data.

Target Mean Game:
Data Distribution and the Mean

INTRODUCE

Project the sketch for viewing by the class. Expect to spend about 15 minutes.

1. Open **Target Mean Game.gsp.** Go to page "Game." Explain, *Today you're going to use Sketchpad to play a game that will challenge your thinking about data distribution and the mean.*

   Let's look at the game board. Point out that the sketch shows the beginning of a line plot. To the left of the line plot is a collection of tiles that students will drag onto the line plot. Because there are no tiles on the line plot when you open the sketch, the "Current Mean" is shown as "undefined."

There are 30 tiles. Creating about 10 to 15 quiz scores for data will be sufficient.

   We need some data to show on the line plot. Suppose I gave a quiz with ten questions and each question was worth one point. What are some scores I might see as I look through students' quizzes? Record responses on the board. If students don't suggest 0, or don't include the same score more than once, introduce these possibilities and add some scores to illustrate them. *Let's represent these data by using the tiles.* One at a time, drag tiles over the numbers representing the recorded quiz scores. (When two or more tiles are dragged to the same number, the tiles should be stacked above each other, not directly onto each other.) Students will note that the model keeps track of the mean, updating it automatically.

   Model pressing *Hide Current Mean* and *Show Current Mean.* Then model pressing *Reset* to remove the tiles.

2. *Now you know how the game board works. Let's talk about the game. It may seem a little backward to you: You will start with a mean and then look for a set of numbers that has that mean.*

   Let's start with a mean of 5.

   Press *Hide Current Mean.* **Can you think of two data points that would have a mean of 5?** Take responses. For the sake of example, we'll use the data points 2 and 8. *Let's drag two tiles to those data points, one at 2 and the other at 8.* Drag the tiles onto the line plot. *Explain how you know the mean is 5.* Here are sample responses.

If students suggest only the standard "add-and-divide" algorithm for finding the mean, pose questions to draw out the thinking expressed in the three responses here. These methods use a "balance" strategy for understanding the mean.

The mean is 5 because 2 and 8 are the same distance from 5. Both tiles are three away from 5.

The mean is 5 because the 2 and 8 balance each other out. One number is three above 5, the other number is three below 5.

If I take 3 away from 8 and give it to the 2, then the numbers become 5 and 5. The mean of 5 and 5 is 5.

The mean is 5 because 2 + 8 = 10, and 10 ÷ 2 = 5.

3. Distribute the worksheet. **Let's learn how to play the Target Mean Game.** Invite two volunteers to the computer to play a round of the game. Follow the rules in worksheet step 2. At the end of player A's turn, the line plot might look like the one shown here. Player B has placed a tile at 9, and player A has followed, placing a tile at 1.

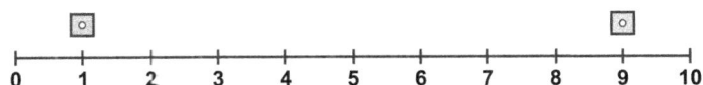

Player A scores two points because there are two tiles on the line plot.

Explain how the players would complete the scorecard.

Scorecard

Target Mean	First Tile	Solution Tile(s)	A's Score	B's Score
5	9	1	2	—

Emphasize that because students score one point for each tile placed in a successful solution, they should look for ways to use more than two tiles. Also be sure students understand that more than one tile may be placed on a number.

Look at the scorecard on the worksheet. Explain that a game consists of three rounds, with each player having a turn in a round. The player with the most points at the end of the game wins.

DEVELOP

Expect students at computers to spend about 30 minutes.

4. Assign students to computers and let them know where to locate **Target Mean Game.gsp.** Make sure each pair has a copy of the worksheet. Tell students to work through step 2 and do the Explore More if they have time.

5. Let pairs work at their own pace. As you circulate, here are some things to notice.

• What strategies do students use to restore the target mean? Here are student explanations that show the use of a balancing strategy.

I need a number that is the same distance from the target mean, but bigger by the same amount.

I need a number that is four less than the mean, because this tile is four more than the target mean.

The target mean is 5. And there's a tile on 1. I'm going to put two tiles on 7—two away from the mean—because that makes four more than the mean, and 1 is four less than the mean.

• To find solutions, some students may rely on the standard "add-and-divide" algorithm. In the game, this strategy is not a very efficient one. As students play more rounds, look to see whether they experiment with a balancing strategy.

• If the target mean is 6 and a tile is placed at 8, the mean can be restored by placing a tile at 4. But this scores only two points, for the two tiles on the line plot. In such a situation, ask, **Is there a way you can score more points?**

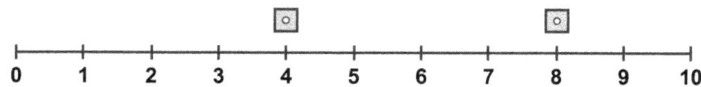

One solution is shown in the following illustration. The 8 is two more than the mean of 6. The two 5's are each one less than the mean of 6. The two 1's "balance out" the two.

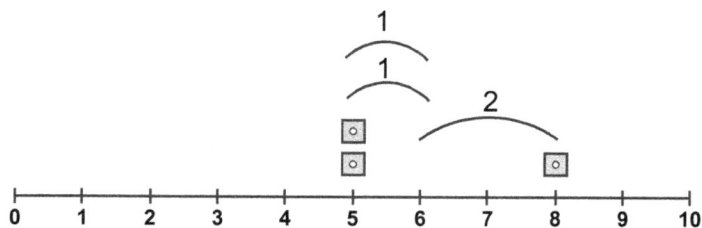

Another way to think about why 5, 5, 6, and 8 have a mean of 6 is to ask, **What happens to the sum of 5, 5, 6, and 8 when I drag the tile at 8 to 6?** [The sum decreases by 2.] **How can we get the sum back to what it was?** [Drag each of the tiles at 5 to 6.] **Now all four tiles are at 6. The average of four 6's is 6.**

- Here are two other strategies that allow a player to score lots of points.

  If the target mean is 6 and a tile is placed at 8, the mean can be restored by placing a tile at 4. Now that the mean is 6 again, a player can place the remainder of the tiles at 6 to keep the mean at 6. This is a worthwhile discovery for students to make, but you might want to prohibit it as an option once it is discovered.

  Another way to keep the mean at 6 is to place many tiles at 4 and the same number of tiles at 8. Again, you and your class should decide whether this is an allowable option.

- The Explore More (worksheet step 3), offers several variations of the game. In version 3a, students must place at least one tile at 0. This addresses a common misconception that adding 0 to a set of data points has no effect on the mean.

  In version 3c, students get to choose the target mean. It's possible that students will create a challenge that is impossible to solve. If the target mean is 0 or 10, for example, and a tile is placed at 1, there is no way to restore the mean.

SUMMARIZE

Project the sketch. Expect to spend about 15 minutes.

6. Gather the class. Students should have their worksheets. Pick one of the target means in worksheet step 2 and ask students to share how they achieved the mean.

   Students will likely be proud of solutions in which they used many tiles. If, for example, the target mean was 2 and a first tile was placed at 7, a student might say: *Seven is five away from 2. That means I needed to balance out the five in the other direction. I placed five tiles at 1. There was no way to solve the problem by adding only one tile.*

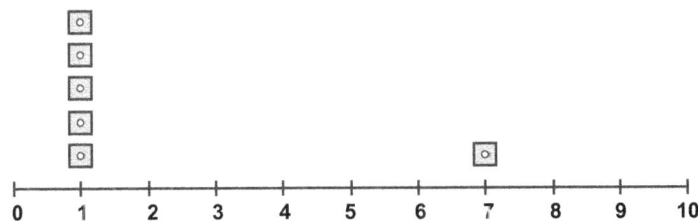

7. When a student has demonstrated a solution that uses five or more tiles, ask, ***How can you find the sum of all the tiles without adding them all up?*** Take responses. In the example from step 6 above, there are seven tiles and their mean is 2. The sum of the tiles is $7 \times 2 = 14$.

8. If students had time to try any of the game variations in Explore More, have them share some of their experiences.

EXTEND

1. ***What other questions about means can you ask?*** Encourage curiosity. Here are sample student queries.

   What if you could only add two tiles to the line plot each time? Could you always get to the target mean?

   How many ways are there of placing tiles to get a certain mean?

   What if we were interested in the median instead of the mean? How would the game work?

2. Encourage students to play versions of the game at a later time.

3. Have students make up their own rules for the game, including scoring systems. They should record the rules, play the game several times, and reflect on whether their rules improve the original game.

ANSWERS

Many correct answers are possible, including, at times, that it is impossible to achieve the mean. Answers will vary depending on the tiles students choose to play.

Target Mean Game

For GSP5 Name:

Use what you know about the mean to play a game.

EXPLORE

1. Open **Target Mean Game.gsp.** Go to page "Game."

2. Play the game. A game consists of three rounds.
   a. Make sure *Hide Current Mean* is pressed to start.
   b. Player B looks at the scorecard to see the target mean for the turn and then drags one tile onto the line plot.
   c. Player A drags one or more tiles onto the line plot to create a data set that has the target mean.
   d. Player A presses *Show Current Mean* to check.
   e. If the current mean matches the target mean, player A gets a point for each tile on the line plot. The solution tiles and score are recorded.
   f. If the current mean does not match the target mean, player A presses *Hide Current Mean* and moves the tiles to try again.
   g. Player A presses *Reset* and *Hide Current Mean* to prepare the board for player B's turn.
   h. The players switch roles. Player A looks at the next target mean on the scorecard and drags one tile onto the line plot. Player B drags one or more tiles to create a data set with that target mean.
   i. When all three rounds have been played, each player's points are totaled. The player with the most points wins.

Scorecard

Round	Target Mean	First Tile	Solution Tile(s)	A's Score	B's Score
1	5				—
1	6			—	
2	7				—
2	4			—	
3	2				—
3	3			—	
			TOTAL POINTS:		

Target Mean Game
continued

For GSP5

EXPLORE MORE

3. Play the game with variations on the original rules.
   a. One of the data points must be 0 (zero).

Target Mean	First Tile	Solution Tile(s)	A's Score	B's Score
5				—
6			—	
7				—
4			—	
2				—
3			—	
		TOTAL POINTS:		

b. The target mean is not a whole number.

Target Mean	First Tile	Solution Tile(s)	A's Score	B's Score
5.5				—
6.5			—	
7.5				—
4.5			—	
2.5				—
3.5			—	
		TOTAL POINTS:		

c. Players take turns choosing the target mean.

Target Mean	First Tile	Solution Tile(s)	A's Score	B's Score
				—
			—	
				—
			—	
				—
			—	
		TOTAL POINTS:		

New York City Title I Elementary School Activities Grades 1–5
© 2012 Key Curriculum Press

Mean Meets the Median:
Measures of Central Tendency

For
GSP5

ACTIVITY NOTES

INTRODUCE

Project the sketch for viewing by the class. Expect to spend about 5 minutes.

1. Open **Mean Meets the Median.gsp** and go to page "Median 5."

2. Explain, *Today you're going to use Sketchpad to compare the behavior of the mean and the median. We'll start with a given data set and explore how changes to the values of that data set affect both the mean and the median. This will help us understand the difference between the two measures of central tendency and when it might be more appropriate to use one or the other. We'll look at questions such as "When is the mean greater than the median?" and "Can the mean and the median ever have the same value?" Before you begin, I'll demonstrate how to change a data set represented in the sketch.*

3. Point out that the vertical orange line represents the median value of the five data points. The line currently goes right through the data point at 4.0. Drag one of the data points along the axis until the median changes value. Drag a different point along the axis until the median changes again.

   - Ask students what they notice about the median value. They might describe it as always being in the middle of the data values, and they might also notice that the vertical line always passes through one of the data points. Make sure such intuitive ideas about the median are shared with the whole class in this introduction portion of the activity.

   - Draw attention to the fact that there are five data points. Later, students will investigate how to calculate the median of a data set that has an even number of data points.

DEVELOP

Expect students at computers to spend about 25 minutes.

4. Assign students to computers and tell them where to find **Mean Meets the Median.gsp.** Distribute the worksheet. Tell students to work through step 5, which wraps up the exploration of the median of five data points.

5. After most students have completed worksheet step 5, ask for examples of data sets with a median of 4.0. Illustrate them in Sketchpad. Ask students for different data sets (these might include data sets where there are two or more values on 4.0, as well as data sets that are either

New York City Title I Elementary School Activities Graces 1–5
© 2012 Key Curriculum Press

373

clustered or spread out). The variety of data sets should show that even though the median value is always the middle value, it does not always look visually centered.

6. Ask students to work through step 13 and do the Explore More question if they have time. Let pairs work at their own pace. As you circulate, here are some things to notice.

   • Make sure students can articulate the difference between the way the median is calculated with five or six data points.

   • Encourage students to drag the different data points to different locations and not just focus on one data point.

   • For steps where students are asked to provide more than one example, encourage them to construct examples that are different. For example, the data sets {1, 2, 3, 4.1, 5}, {1, 2.1, 3, 4, 5}, and {1, 2, 3, 4, 5} are similar examples with medians of 3. However, {0, 0, 3, 3, 10} is quite different because it contains both repeated and extreme data points.

   • Encouraging students to see what their data sets have in common might help them construct different data sets. (Sometimes it's hard for them to think outside a particular strategy they have developed to generate new data sets.)

   • In worksheet step 13, you might ask students to try some sample data values so that they see that the waiting times for the patients who stayed all morning are much longer than for the other patients. Therefore the corresponding data points might be considered outliers.

SUMMARIZE

Project the sketch. Expect to spend about 15 minutes.

7. Gather the class. Students should have their worksheets with them. Begin the discussion by opening **Mean Meets the Median.gsp** and use it to support the class discussion.

8. Ask students to offer explanations for worksheet step 8. If students do not explicitly mention the words *odd* and *even*, probe their understanding by asking them what might happen in the case of data sets with 11 or 12 data points. They should be able to generalize that to find the median of an odd data set one simply locates the middle number, and to find the median of an even data set one must find the mean of the two middle numbers.

9. Now discuss the question in step 10, and ask students to give examples of three different ways to make the mean and median equal to each other. One might be to change the data point representing the median value, whereas another might be to increase or decrease one of the other values to push the mean closer to the median. Help students notice that there are many different data sets that will have the same mean and median, so that these two measures of central tendency do not completely describe a data set.

10. Students have now worked quite a bit on constructing different sets of data. *Now we'll try to characterize the major differences between the mean and the median. Which measure of central tendency is more sensitive? Which might be more useful in different circumstances? Using the sketch, on page "Mean and Median" set the data values to 5.1, 5.4, 5.9, 5.9, and 6.2. Imagine these are the heights of five people lined up to go through security at the airport. The mean is lower than the median. Now suppose that one of the people lined up was a toddler, who measured 2.3. This new data value can be thought of as an outlier. How would the mean and median change?* Drag one of the first values to 2.3. Now the mean will be much smaller than the median. This illustrates the way in which outliers can affect the mean without affecting the median at all.

11. Continue on to worksheet step 13. Ask for volunteers to describe their answers. Students should be able to recognize that the waiting times that are outliers influence the decision about hiring a new doctor, so the median might be a better measure. *Are there other situations you can think of where you would want to consider the outliers?* Encourage students to propose a variety of situations.

EXTEND

Suppose a teacher has just marked a mathematics test she gave to her students. Can you find a reason why she might prefer to know the mean of the test scores? Can you find a reason why she might want to know the median? Can you find a reason why she might want to know both? What other pieces of information might be useful to her? Encourage students to play with specific examples if they need to. If we want to know whether a test was too hard for most students, then the mean is helpful because it is sensitive to extreme values (in this case, to very low test scores).

However, the median might be a good measure if we don't care about extreme values (maybe some students were away, so we don't want their marks of 0 to affect the measure). It might also be helpful to know the range of the values. If the range is relatively wide, we might conclude that the test was not a very good one, because some students found it much too easy and others, much too hard.

You might also ask what data sets with equal means and medians have in common, and whether there are other measures of central tendency that have the advantages of each (for instance, dropping the highest and lowest values and finding the mean of the remaining values, as they do in the Olympics).

ANSWERS

1. Answers will vary. All data sets should include the value 4.0.

2. The median won't change as long as the data value remains greater than 4.0.

3. Once the data value becomes smaller than 4.0, the median will change because 4.0 is no longer the middle value.

4. Answers will vary. The data set should include two points with the same value as the median.

5. The median is the middle value when the data points are placed in order, but sometimes the middle value will not be visually centered among the data points.

6. Answers will vary.

7. Answers will vary. Some students might start by placing two data points on 7.5 and then two data points on each side of 7.5. Others might start with two data points whose average is 7.5 and then place two data points on each side of 7.5.

8. In step 1, with five data points the median is the middle value, but now the median is the average of the two middle values.

10. Answers will vary.

11. Answers will vary. Increasing the largest data value will not affect the median, but it will increase the mean.

12. Only the mean. Increasing the value of one data point does not change the middle value, but it does change the sum, and therefore the mean, of the values.

13. It's probably better to use the median because we don't want to include the extra waiting time of the patients who arrive early.

14. Answers will vary. The sum of the values of the two data points remains constant, so the mean is unaffected.

Mean Meets the Median

For GSP5 Name:

In this activity you'll investigate some properties of the median and compare its behavior to that of the mean.

EXPLORE

1. Open **Mean Meets the Median.gsp.** Go to page "Median 5." You should see data points at 1.0, 2.0, 4.0, 7.0 and 8.0, and a thick vertical orange line through 4.0 that represents the median value. Drag the data points. Write down different data sets that have a median of 4.0. What do all these data sets have in common besides having the same median?

2. Press *Reset.* Drag the data point at 8.0 to different locations. Describe all the values that data point can have without affecting the median value.

3. Press *Reset.* Drag the data point at 8.0 so that the median value changes. Why did it change?

4. Create a data set that has two data points on one side of the orange line and only one data point on the other side. Describe what's special about your data set.

5. Some people say that the median value is always in the middle. Why might this be a misleading way to characterize the median?

6. Go to page "Median 6." Drag the data points. Write down three data sets that all have a median of 4.0.

7. Describe a strategy for creating a data set that has a median of 7.5.

New York City Title I Elementary School Activities Grades 1–5
© 2012 Key Curriculum Press

8. In step 1, you found that all your data sets had to contain the value 4.0. Is this still true? Explain.

9. Go to page "Mean and Median." Try to predict the value of the mean for the given data set. Then press *Show Mean.*

10. Find three different ways of changing the data points so that the mean and the median are equal to each other.

11. Find a data set in which the mean and median values are the same. Predict what will happen to each value if you increase your largest data value. Verify your prediction.

12. In many cases extremely small or large data values are called *outliers.* Will the mean or the median be more affected by outliers? Explain.

13. A doctor's office wanted to find out how long patients had to wait in order to see whether they needed to hire another doctor. After gathering their "waiting time" data, they learned that some patients had been dropped off first thing in the morning, even though their appointments weren't until late morning or early afternoon. Which measure of central tendency do you think would be more useful to the doctor's office?

EXPLORE MORE

14. Go to page "Median 6." Press *Show Mean.* Every time you move one of the data points, the mean changes. However, if you could move two data points at once, you could probably keep both the mean and the median the same. Press *Reset.* Select the data points 1.0 and 8.0 and choose **Edit | Action Buttons | Animation.** Set point *C* to animate in a forward direction and point *E* to animate in a backward direction. Press the button you just created to verify that your mean value stays the same. Explain why this works.

www.ingramcontent.com/pod-product-compliance
Lightning Source LLC
Chambersburg PA
CBHW080707220326
41598CB00033B/5337